Thin Films

Edited by Alicia Esther Ares

Published in London, United Kingdom

IntechOpen

Supporting open minds since 2005

Thin Films
http://dx.doi.org/10.5772/intechopen.87838
Edited by Alicia Esther Ares

Contributors
Chinecherem Nkele, Fabian Ezema, Ludmila Otilia Cinteză, Maria Antonia Tănase, Edwin Acosta, Abtisam Al-Masoodi, Boon Tong Goh, Ahmed H. H. Al-Masoodi, Wan Haliza Binti Abd Majid, Manish Kumar, Sunita Rani, Chirag Saharan, Judith Chebwogen, Christopher Mkirema Maghanga, Dwight Acosta-Najarro, Ismael A. Garduño-Wilches, Arturo Maldonado, Maria de la Luz Olvera, Sabrina Roguai, Abdelkader Djelloul, Marcos Antonio Santana Andrade Junior, Lucia Helena Mascaro, Hugo Leandro Sousa Dos Santos, Mileny Dos Santos Araujo, Arthur Corrado Salomão, Kashif Tufail Chaudhary, Mohamed Masmoudi

First published in London, United Kingdom, 2021 by IntechOpen
IntechOpen is the global imprint of INTECHOPEN LIMITED, registered in England and Wales, registration number: 11086078, 5 Princes Gate Court, London, SW7 2QJ, United Kingdom
Printed in Croatia

British Library Cataloguing-in-Publication Data
A catalogue record for this book is available from the British Library

Additional hard and PDF copies can be obtained from orders@intechopen.com

Thin Films
Edited by Alicia Esther Ares
p. cm.
Print ISBN 978-1-83881-986-6
Online ISBN 978-1-83881-993-4
eBook (PDF) ISBN 978-1-83881-994-1

Meet the editor

Alicia Esther Ares obtained a degree in Chemical Engineering from the National University of Misiones, Argentina in 1992, and an MSc in Science and Technology of Materials and Ph.D. in Science and Technology, Materials Mention, both from the Institute of Technology "Professor Jorge A. Sabato," National Atomic Energy Commission and National University of General San Martin, Buenos Aires, Argentina in 1997 and 2000, respectively. She completed postdoctoral stays at several institutions, including Faculdade de Engenharía Mecânica, Departamento de Engenharía de Materiais, Universidade Estadual de Campinas, São Paulo, Brazil; Department of Materials Science and Engineering, University of Florida, Gainesville, Florida, United States (2002–2003); and Faculty of Sciences, National University of Misiones (2003–2004). She has been a researcher with the National Scientific and Technical Research Council (CONICET) since 2004, and Professor of Materials Science in the Chemical Engineering Department, School of Sciences (FCEQyN), National University of Misiones. She has more than thirty years of teaching experience both at the undergraduate and graduate levels, and is a member of various societies, including the Minerals, Metals and Materials Society (TMS, USA), and International American Society for Testing Materials, among others.

Contents

Preface XI

Section 1
Properties and Applications 1

Chapter 1 3
Thin Films/Properties and Applications
by Edwin Acosta

Section 2
Characterization 27

Chapter 2 29
Infrared Characterization and Electrochemical Study of Silanes Grafted into Surface of Copper
by Mohamed Masmoudi

Chapter 3 43
Fabrication and Characterization of Cobalt-Pigmented Anodized Zinc for Photocatalytic Application
by Judith Chebwogen and Christopher Mkirema Maghanga

Section 3
Nanoparticles 55

Chapter 4 57
Diverse Synthesis and Characterization Techniques of Nanoparticles
by Agnes Chinecherem Nkele and Fabian I. Ezema

Chapter 5 69
Roles of Cobalt Doping on Structural and Optical of ZnO Thin Films by Ultrasonic Spray Pyrolysis
by Sabrina Roguai and Abdelkader Djelloul

Chapter 6 89
Multifunctional ZnO Nanoparticle: Based Coatings for Cultural Heritage Preventive Conservation
by Ludmila Otilia Cinteză and Maria Antonia Tănase

Chapter 7 109
Deposition of Silver Nanoparticles on Indium Tin Oxide Substrates by Plasma-Assisted Hot-Filament Evaporation
by Abtisam Hasan Hamood Al-Masoodi, Boon Tong Goh, Ahmed H.H. Al-Masoodi and Wan Haliza Binti Abd Majid

Section 4
Photovoltaic Thin Films 129

Chapter 8 131
Solution-Processed Chalcogenide Photovoltaic Thin Films
by Marcos Antonio Santana Andrade Junior, Hugo Leandro Sousa dos Santos, Mileny dos Santos Araujo, Arthur Corrado Salomão and Lucia Helena Mascaro

Section 5
VO_2 Polymorphs 145

Chapter 9 147
Thin Film Stabilization of Different VO_2 Polymorphs
by Manish Kumar, Chirag Saharan and Sunita Rani

Section 6
Thin Film Deposition 163

Chapter 10 165
Thin Film Deposition: Solution Based Approach
by Kashif Tufail Chaudhary

Section 7
Sol-Gel Dip Coating Method 177

Chapter 11 179
TiO_2 Thin Films Produced by Ultrasonic-Agitation Assisted Sol-Gel for Propane Gas Sensor
by Dwight Acosta-Najarro, Ismael A. Garduño-Wilches, Maria de la Luz Olvera and Arturo Maldonado

Preface

Thin films are thin material layers ranging from fractions of a nanometer to several micrometers in thickness. The properties of material differ significantly when analyzed in the form of thin films. Thin film technologies make use of the fact that the properties can be controlled by the thickness parameter. Thin films are formed by deposition generally, using chemical or physical methods. Thin films, both crystalline and amorphous, have enormous importance in the age of high technology. Thin film technology is a relatively new and growing field in the physical and chemical sciences, which is a confluence of materials science, applied chemistry, applied physics, surface science, and others.

The properties of thin films are extremely sensitive to the method of preparation. Several techniques have been developed and are discussed in this book.

Section I: Properties and Applications

Chapter 1, "Thin Films/Properties and Applications" by Edwin Acosta, covers the fundamentals of thin film technology, including deposition techniques, structure and morphology, film properties, characterization techniques, and applications.

Section II: Characterization

Chapter 2, "Infrared Characterization and Electrochemical Study of Silanes Grafted into Surface of Copper" by Mashmoudi Mohamed, investigates the adsorption of γ-MPS and γ-APS onto pure copper at different dried conditions and studies the possibility of using these two coupling agents' treatment on copper to enhance its corrosion protective performance.

Chapter 3, "Fabrication and Characterization of Cobalt-Pigmented Anodized Zinc for Photocatalytic Application" by Judith Chebwogen and Christopher Mkirema Maghanga, shows that ZnO film pigmented with cobalt for 20 seconds was found to be the most photocatalytic with a rate constant of 0.0317hr-1. Thus, it had the optimum cobalt concentration for photocatalytic water treatment and can be applied in small-scale water purification.

Section III: Nanoparticles

Chapter 4, "Diverse Synthesis and Characterization Techniques of Nanoparticles" by Agnes Chinecherem Nkele and Fabian I. Ezema, provides insight into the knowledge of nanoparticles, their classifications, parameters affecting their efficient performance, and synthesis and characterization techniques.

Chapter 5, "Roles of Cobalt Doping on Structural and Optical of ZnO Thin Films by Ultrasonic Spray Pyrolysis" by Sabrina Roguai and Abdelkader Djelloul, shows how the authors obtained thin layers of ZnO, Zn1-xCoxO deposited by the ultrasonic pyrolysis spray technique at a temperature of 450°C lasting 30 min. The authors report the structural and optical properties of these thin films. Particular attention is given to the theoretical methods used for determining the dispersion parameters of the films using only a single transmission spectrum.

Chapter 6, "Multifunctional ZnO Nanoparticle: Based Coatings for Cultural Heritage Preventive Conservation" by Ludmila Otilia Cinteză and Maria Antonia Tănase, describes recent progress in synthesis and characterization of hybrid material with ZnO nanoparticles and their use as functional coatings for various substrates.

Chapter 7, "Deposition of Silver Nanoparticles on Indium Tin Oxide Substrates by Plasma-Assisted Hot-Filament Evaporation" by Abtisam Hasan Hamood Al-Masoodi, Boon Tong Goh, Ahmed H. H. Al-Masoodi, and Wan Haliza Binti Abd Majid, discusses the PAHFE technique for the deposition of Ag nanoparticles on ITO and influences of the experimental parameters on the physical and optical properties. It also examines the electronic structure of the deposited Ag nanoparticles on ITO.

Section IV: Photovoltaic Thin Films

Chapter 8, "Solution-Processed Chalcogenide Photovoltaic Thin Films" by Marcos Antonio Santana Andrade Junior, Hugo Leandro Sousa dos Santos, Mileny dos Santos Araujo, Arthur Corrado Salomão and Lucia Helena Mascaro, discusses the two most reported techniques used to prepare world-record efficiencies of solution-processed chalcogenide solar cells: spin-coating and spray-coating.

Section V: VO_2 Polymorphs

Chapter 9, "Thin Film Stabilization of Different VO_2 Polymorphs" by Manish Kumar, Chirag Saharan and Sunita Rani, focuses on the stabilization of thin film of six main VO_2 polymorphs: VO_2 (M1), VO_2 (M_2), VO_2(R), VO_2(T), VO_2(A) and VO_2(B). It should be noted that VO_2 polymorphs VO_2 (M_3), VO_2(P), VO_2(C), and VO_2(D) have also been studied in bulk and nanostructure form and reports are missing on thin film stabilization of these phases.

Section VI: Thin Film Deposition

Chapter 10, "Thin Film Deposition: Solution Based Approach" by Chaudhary Kashif, presents the details of solution-based routes to deposit thin films, including spray coating, dip coating, spin coating, and inkjet printing processing. It discusses the contribution of different experimental parameters such as solution viscosity, surface tension, droplet size, substrate material and temperature, and nature of the solution.

Section VII: Sol-Gel Dip Coating Method

Chapter 11, "TiO_2 Thin Films Produced by Ultrasonic-Agitation Assisted Sol-Gel for Propane Gas Sensor" by Dwight Acosta, Ismael A. Garduño-Wilches, Maria de la Luz Olvera, and Arturo Maldonado, describes TiO_2 films obtained by the dip-coating method using ultrasonic assistance during the crystal growth stage. The film thickness and work temperature are used as variables to explore the response to propane gas at different concentrations. The surface morphology is studied for different immersion cycles and the crystalline structure is analyzed by x-ray diffraction.

Alicia Esther Ares
Facultad de Ciencias Exactas,
Universidad Nacional de Misiones (UnaM),
Químicas y Naturales (FCEQyN),
Instituto de Materiales de Misiones (IMAM),
Programa de Materiales y Fisicoquímica (ProMyF),
Posadas, Misiones, Argentina

Section 1

Properties and Applications

Chapter 1

Thin Films/Properties and Applications

Edwin Acosta

Abstract

Since its discovery in early times, thin films rapidly found industrial applications such as in decorative and optics purposes. With the evolution of thin film technology, supported by the development of vacuum technology and electric power facilities, the range of applications has increased at a level that nowadays almost every industrial sector make use of them to provide specific physical and chemical properties to the surface of bulk materials. The possibility to tailor the film properties through the variation of the microstructure via the deposition parameters adopted in a specific deposition technique has permitted their entrance from the simplest like protective coatings against wear and corrosion to the most technological advanced applications such as microelectronics and biomedicine, recently. In spite of such impressive progress, the connection among all steps of the thin film production, namely deposition parameters-morphology and properties, is not fully accurate. Among other reasons, the lack of characterization techniques suitable for probing films with thickness less than a single atomic layer, along with a lack of understanding of the physics have impeded the elaboration of sophisticated models for a precise prediction of film properties. Furthermore, there remain some difficulties related to the large scale production and a relative high cost for the deposition of advanced structures, i.e. quantum wells and wires. Once these barriers are overcome, thin film technology will become more competitive for advanced technological applications.

Keywords: thin films, deposition techniques, characterization, properties, inorganic and organic, applications, structure and morphology, challenges

1. Introduction

This chapter covers the fundamentals of the thin film technology, including deposition techniques, structure and morphology, film properties, characterization techniques and applications. Due to the exceptional diversity of thin films and the large range of applications it is complicated to cover in detail all the areas, however, as many thin films share a number of features in all steps of the production process, characterization, and even applications, it is possible to treat them with a general perspective.

The most widely employed deposition techniques, namely physical vapor deposition (PVD), and chemical vapor deposition (CVD), for thin film production are described with emphasis in the principle, advantages, disadvantages and examples of the common thin film structures produced. Furthermore, the various types of microstructures and the morphological features as a result of the deposition technique and parameters are reported. The growth mechanism is described considering the conditions for the formation of amorphous, polycrystalline or epitaxial structure, and the resulting morphology is explained based on the temperature-dependence zone models. Subsequently, the mechanical, electrical and optical properties are reviewed, stressing the connection with morphological features such as film thickness, material phase, defects, roughness, grain boundaries etc. The principal characterization techniques used for the characterization of chemical composition, morphology, stress, and electrical conductivity are described along with their principle. In addition, the specific information that can be accessed through the data collection is stated. The most technological relevant areas of application are then outlined along with the type of films mostly employed and. Finally, the main challenges for the thin film technology to progress in various areas are discussed.

2. Deposition methods

Thin-films are in general developed to provide special properties, i.e. electrical, optical, mechanical, chemical, that satisfy the needs for specific applications. The desired properties are determined by the resulting film structure, which strongly dependents on the selected deposition method, film material, and substrate. In line with the wide range of applications of thin films, a number of deposition methods have been developed/improved to optimize the film properties, of which, the most commonly employed are described in this section. Broadly speaking, thin-film production can be realized based on two technological groups, namely physical and chemical deposition methods.

2.1 Physical deposition methods (PVD)

Physical deposition methods are usually referred as to physical vapor deposition methods (PVD) because the process entails the generation of vapor. PVD essentially consists in removing growth species from a source or target material via evaporation, then this vapor is transported to the substrate surface, and eventually it solidifies in the surface, forming the film. The evaporation is generally carried out under a reduced pressure chamber to avoid impurities in the film formation which are produced due to collisions between vapor particles and residual gas particles in their displacement from the source to the substrate surface. PVD techniques are known to offer a number of advantages, including the deposition of almost any material, high reproducibility of film properties, the use of a large range of substrate materials, the possibility of tailoring the film properties through modification of deposition parameters in single element deposition, and obtaining films with high purity. On the other side, among the main disadvantages are the use of sophisticated and costly monitoring systems for the control of the deposition rate and film thickness, and the mismatch between the composition of the deposited film and the composition of the evaporant in the case of alloys and compounds.

PVD techniques can be classified according to the method employed in the generation of vapor. The most common PVD techniques are vacuum-based evaporation and its heating versions, sputtering, laser ablation, cluster beam and ion pattering, of which, only the most important will be described in detail in the present section. For further detail on the versions of PVD techniques, the reader is referred to [1–3].

2.1.1 Vacuum evaporation

It is among the most popular PVD techniques due to its simplicity in operation and high deposition rate. This technique uses heating sources to evaporate the deposition material onto the substrate surface where it condenses forming a thin film, all within a vacuum chamber. This technique is suitable for deposition of elements or compounds at temperatures below 2000 K [4]. According to the method used to evaporate the target material, this technique can be subdivided into resistance-heated evaporation, and electron-beam evaporation.

In *resistance-heated evaporation* the target material is deposited in one of the multiple configurations of the evaporation source including coil, boats, and baskets. Due to the high temperature required, they are commonly fabricated from refractory metals with high melting points, such as tungsten, molybdenum, and tantalum although stainless steel can also be used in cases where the target material has a low evaporation temperature. The essential condition for the proper selection of an evaporation source is that its evaporation point does not have to be reached at the operation temperature. Once selected an appropriate evaporation source the vacuum chamber needs to reach pressures lower than 10^{-5} mbar to optimize the sample coverage during deposition, and provide a high purity of the film. The evaporation of the target material is carried out by the heating of the source through which a high electric current is forced to pass. The deposition rate is controlled with the source temperature due to its direct relationship with the vapor pressure of the target material. Among the main disadvantages of this technique are the limited upper temperature (2000 K) which constrains the use of materials suitable for evaporation, a limited film thickness, and the possibility of contamination related to the heater filament.

In *electron-beam evaporation* the target material is deposited in a crucible which is design to match the heating and power density of the electron-beam. To avoid contamination of the target material in the evaporation stage, the crucible should not be prone to evaporation or erosion at high temperatures. Accordingly, the most widely used crucibles are the water-cooled copper, and ceramic hearth. With these crucibles, which have high melting points (3000–4000) K, it is possible to reach higher temperatures than those achieved for resistance-heated evaporation, and thus, it allows the evaporation of refractory material, reactive materials like titanium and aluminum, dielectrics (SiO_2), boron, carbon, silicon among others. The evaporation process takes place under vacuum through the incidence of an intensive beam of electrons, emitted by a thermionic filament, which is accelerated towards the target material by an electric field. These energetic electrons collide in the surface of the target material generating a local melting after having being deflected by a magnetic field usually with an angle of 180° or 270°. The deposition rate achieved is high, in the order of 1000 nm/min, relying on the target material, and the distance between the source and substrate. One of the main disadvantages of this technique is the generation of ionized radiation (X-rays) which can penetrate the film producing damage.

Molecular-beam epitaxy (MBE) is the most advance solution for the deposition of compounds and alloys due to its simultaneous control over the evaporation rates of the different constituents, with high precision. This technique enables the epitaxial grow of thin films under ultra-high vacuum (UHV) conditions onto a hot single-crystalline substrate. While the UHV prevents contamination of the film from impurities present in the growth environment producing a high-purity film with an improved morphology, the substrate temperature promotes the diffusion of adsorbed species on its surface to achieve an epitaxial growth. The growing species are provided by the atomic or molecular beams generated by the heating of a convenient source material which can be either elements or compounds, and that are normally contained in crucibles of high purity. The proper selection of crucibles is fundamental to assure the purity of the molecular beam; the most widely employed crucibles are fabricated based on graphite, pyrotilic boron nitride for high temperatures, and also quartz and stainless steel are for low melting point materials.

One of the most important features of this technique is its ability to accurately control the composition and doping profile in the growth direction. This can be realized thanks to the UHV that enables the film growth to occur in the molecular regime. In this regime the atoms and molecules do not collide in their way to the substrate because the mean free path is larger than the distance from the crucible to the substrate. This feature allows to modify as desired the composition of the feeding phase, blocking abruptly one or more streams of atoms being either a constituent atom or a dopant element with the help of mechanical shutters. Regarding that this blocking process can be carried out in very short times, and the low deposition rate from MBE, it is possible to achieve extremely thin thicknesses between the layers of different composition and/or doping. Advanced nanostructures, such as quantum-wells, superlattices and quantum-dots have been successfully fabricated with this deposition technique [5–7].

2.1.2 Sputtering

Sputtering essentially consists in the bombardment of the target material with energetic particles to dislodge atoms from its surface which travel through the plasma to eventually condensate onto the substrate. Three sputtering techniques are the most employed for thin film growth, including DC diode, RF-diode, and magnetron diode.

In the DC diode sputtering, the bottom electrode, called cathode, contains the target material to be deposited while the top electrode holds the substrate. An inert gas, i.e. Ar, Ne, Kr, or Xe [8] is fed into the sputtering reactor at a reduced pressure. A plasma is then formed by the application of a voltage between the cathode and anode with the inert gas inside. The electrons emitted from the cathode are accelerated towards the anode and in their way ionize the gas molecules producing positively charged ions which accelerate towards the cathode, establishing a discharge. The glow discharge can be made self-sustained if appropriate conditions of gas pressure, voltage and distance between electrodes are adopted. If that occurs, a continuous bombardment of positive energetic ions against the cathode is established promoting its surface sputtering, and the subsequent condensation of a thin film onto the substrate. It should be noted that the target material in this system is necessarily a metal as the glow discharge can be maintained only between metals. Thus, deposition of insulators might not be possible using this technique. The main disadvantages of this technique are the low deposition rates, and the absence of self-sustained glow discharge at very low pressures.

Radio-frequency (RF) sputtering offers a solution for the deposition of insulating materials by preventing the accumulation of positive ions in the front side of the insulator to maintain the discharge. A continuous sputtering process of the insulator target is set by the application of an alternating signal to the cathode with a frequency corresponding to radio-frequency (13.56 MHz) [9]. In this way, a larger number of electrons arrive at the surface target during the positive half-cycle which overcompensate the number of ions that accumulate during the negative half-cycle, giving the target a net negative charge. This self-biased charge process is possible due to the higher mobility of electrons with respect to the positive ions in the plasma. Apart from insulators, this technique has successfully been adopted for the deposition of metals and semiconductors, however, the sputtering rate remains lower than that of the vacuum deposition described above.

The *magnetron sputtering* technique permits to improve the sputtering rate with respect to the previous versions through the optimisation of the ionization of the sputtering gas molecules [4]. In this system, a magnetic field is applied between the cathode (target material) and anode (substrate) in addition to the electrical field where they both make the electrons to change their paths within the discharge from linear to spiral. This longer path followed by electrons in their travel from the target to the anode increases the time they are in contact with the gas molecules, raising the probability of ionization. Also, the fact that electrons remain a longer time in the plasma produces an increase of the current density at the cathode side, which in turn, raises the sputtering rate of the target material. This higher effective gas ionization allows to reduce the sputtering gas pressure to maintain the discharge, which significantly reduces the collisions of sputtered atoms within the plasma in their travel towards the substrate, increasing the deposition rate. A columnar microstructure is usually obtained with this technique with defects concentrated in between the columns. Silicon nitride films are often produced with this technique [10], but in combination with RF sputtering, it is possible to produce films based on Ti, Cr, Fe, Mo, Ag, Cu, among others [11].

2.1.3 Laser ablation

Laser ablation is based on a similar configuration as the previous techniques, i.e. usually an evacuated chamber with a target material to be evaporated, a substrate placed parallel to the target where the film condensates, and in this case, an additional high-power pulsed laser placed outside the deposition chamber which emits the energy for inducing the ablation of the target material [12]. This technique is broadly employed in the deposition of alloys, compounds, polymers, semiconductors, and multilayers due to its excellent stoichiometry transfer from the target to the film. Oxide thin films can also be deposited if oxygen is introduced in the chamber as a background gas. The irradiation from the laser power, i.e. KrF (248 nm), is focused on the target producing a rapid local heating until reaching the melting point, and eventually producing evaporation that will be deposited in the substrate. The laser ablation process and the quality of the sample are affected by a number of parameters, including the characteristics of the target material, deposition conditions, laser parameters, substrate temperature. Although this technique is widely recognized for its diverse and fast applicability, the actual ablation process is yet to be fully understood because the material ejection is not produced solely by a thermal process but also a photochemical is likely present.

2.2 Chemical deposition methods (CVD)

CVD is a deposition method where a volatile compound of a pre-established substance is introduced into a reactor, usually along with an inert gas, to induce a chemical reaction which produces a solid thin film onto a substrate at an elevated temperature. In this technique, unlike PVD, the reaction does not have to be produced under vacuum conditions. Due to its versatility to work with a broad range of reactants and precursors, this technique enables the deposition of a variety of structures, including metal alloys and compound semiconductors with an excellent control of purity and doping (stoichiometric film) [13]. Compared to PVD, this technique offers higher deposition rates, better conformance in rough substrates, easy deposition onto complex surfaces, and high throughput. However, some disadvantages, such as the use of high substrate temperatures, and the toxicity and flammability of the reactive gases have prevented it from being used in low-scaled developments, but is well justified in applications where high-throughput is required, i.e. semiconductor industry. The CVD processes can be classified based on the type of source employed to initiate the chemical reaction, the range of pressure under which the deposition is carried out, and the type of reactant used. The most established CVD method are described below.

The thermal activated CVD is the most conventional method where the thermal energy produced inside the reactor triggers the chemical reactions. Two variants are noticeable within this process in relation to the pressure range under which the deposition materializes, namely atmospheric pressure CVD (APCVD), and low pressure CVD (LPCVD) [14]. Both variants essentially comprise the same chemical reactions which overall consist in the creation of vapor from reactants; then these vapor species are directed into the reactor, where depending on the deposition parameters, homogeneous chemical reactions take place in the gas phase while heterogeneous reactions occur near the substrate surface. Finally, the crystallization of a solid film occurs in the substrate surface. An alteration of the rate-limiting process occurs when the pressure is reduced in LPCVD about 1000-fold with respect to the atmospheric pressure CVD [13]. This reduced pressure substantially increases the rate of mass transport of reactants overpassing the rate of reactions in the substrate surface. This makes the kinetics in the substrate surface to be the rate-limiting step in the deposition process. In the APCVD instead, the mass transport of reactants is lower than the rate of reactions in the substrate surface, thus making the process limited by the diffusion of reactants. The introduction of unwanted impurities is reduced in LPCVD with respect to the APCVD. Moreover, high uniform films with a higher throughput can be produced for commercial applications using LPCVD, however, a disadvantage in this variant is the still high temperature needed for deposition. Polysilicon and silicon nitride films are among the films produced with this technique [15].

Plasma-enhanced CVD (PECVD) is one of the most widely used variations of CVD because it provides an alternative for deposition at lower temperatures using organic, inorganic and inert precursors [4]. In this method a plasma energy, in addition to the thermal energy, is incorporated to improve the dissociation of the reactive gases. The plasma is created when an energy, usually in the form of an electric field, is introduced into the reactor which contains the reacting gases in the space between two electrodes. Then, complex chemical reactions occur in the plasma under reduced pressures which produce energetic ions and radicals that travel towards the substrate bombarding its surface and promoting reactions that give rise to a solid film. Unlike APCDV or LPCDV, this variant allows deposition of conformal films at lower temperatures (200–400 K) [13]. This deposition method has been enhanced through the excitation of plasma by radio frequency field (RF = 13.56 MHz), and microwave frequency field (2.45 GHz) which draws on the effect of electron cyclotron resonance (ECR) to reduce the pressure. The higher frequency of the latter allows for a higher-energy and a higher concentration of electrons in the plasma leading to a boost in the degree of ionization up to ~1000 times than that realized with RF field [16]. PECVD is an established deposition method for research and industrial production, including antireflective coatings, microelectronics, photovoltaics, and transistors.

Metalorganic CVD (MOCVD) basically follows the same process as the CVD where volatile metalorganic compounds are used as precursors instead of inorganic ones. This technique allows the deposition of a broad range of materials with an amorphous, polycrystalline and epitaxial microstructure. The metalorganic compounds are decomposed through pyrolysis reactions at low temperatures which permits to carry out the film deposition at lower temperatures than in thermal CVD.

Both deposition temperature and pressure play an essential role in determining the rate-limiting process. Deposition is usually performed in the range 550–1150 K, noticing a kinetic reaction limited behavior for temperatures below 500 K, and a diffusion-rate limited for temperatures above 800 K [13]. At low pressures (<1 kPa) the reactions are kinetic limited while above that threshold the reactions become diffusion-rate limited. This technique is attractive because the gas flow rate and the partial pressure of the precursors can be controlled allowing the fabrication of films with the right stoichiometry at high deposition rates. Among the disadvantages are the relative high cost of metalorganic compounds along with the difficulty to obtain highly purity version with minimum oxygen content to fabricate high quality semiconductors. Both semiconductor and superconductor thin films have been deposited through this technique [16]. A list of the most used metalorganic precursors for the deposition of semiconductor, metallic, and dielectric films is presented in Ref. [13].

3. Thin films structure

Different and complex chemical reaction occur during the deposition process depending on the technique employed and deposition parameters, such as substrate temperature, deposition rate, pressure, and alignment of vapor stream with substrate, which give rise to a variety of microstructures. The resulting microstructure in turn defines the physical and chemical properties of the film, which means that an appropriate management of these variables is essential to obtain tailored film properties. The resulting microstructure can be either amorphous, polycrystalline or epitaxial, which are briefly described below.

Amorphous thin films are essentially short-range order structures derived from deviations in the bond length and bond angle from a perfect crystalline lattice [17]. Overall, in most materials the growth of amorphous films take place at low substrate temperatures where the mobility of adatoms (adsorbed atoms) at the substrate surface is very limited. With a low temperature the adatoms approach to a thermal equilibrium with the substrate limiting the energy available for diffusion through its surface. Instead, these nearly immobile adatoms are incorporated almost at the point of strike with the substrate surface. High deposition rate is another parameter to induce amorphous growth because it prevents adatoms from migrating to more energetic sites to reach equilibrium due to the limited time for diffusion, and they are incorporated into the film structure almost at the point of strike with the substrate surface. Some developments have also reported the formation of this disordered microstructure by the incorporation of certain gases, i.e. oxygen, nitrogen, that inhibit the growth of crystallites during the deposition process [13]. Both PVD and CVD techniques are suitable for deposition of amorphous structures. Deposition of amorphous thin films are required in a number of applications, such as solar cells, transistors, optoelectronics, dielectric films, etc.

Polycrystalline thin films consist in a large number of nano/micro crystallites with different orientations separated by grain boundaries. The crystallite size is mainly determined by the deposition parameters, i.e. deposition temperature, and deposition rate. A higher deposition temperature than that applied to obtain amorphous structures may lead to the formation of polycrystalline thin films. When adatoms are not in thermal equilibrium with the substrate due to a high substrate temperature, they have enough energy to continue diffusing in the substrate surface until adhering to an existing island or giving rise to new islands. These islands do not become thermodynamically stable until their size reach a nucleation threshold. The stable islands continue growing until saturation, and then, a coalescence process initiates between islands giving rise to the formation of a polycrystalline layer [4]. As adatom-diffusion is a temperature-dependent process, the crystallite size is expected to increase in line with the increase of substrate temperature. Another driver for the crystallite size is the film thickness. While the crystallite growth in the lateral direction is limited by the coalescence process, in the cross plane direction, the growth is limited by the film thickness. The typical microstructure of this material is of great importance for some applications where the scattering of carriers and phonons need to be controlled, i.e. thermoelectrics [18, 19].

Epitaxial thin films consist in a solid crystalline film deposited onto a substrate surface with a nearly perfect lattice structure, whose crystal orientation is aligned with the crystallographic orientation of the substrate surface. Depending on the nature of substrate employed, the epitaxial growth can be divided into homo and hetero-epitaxial, with the former referring to the growth of a film onto a substrate of the same material, and the latter, onto a substrate of different material. Various deposition techniques have been developed to enable the deposition of epitaxial films, including MBE and vapor-phase epitaxy; they are mostly used in the semiconductor industry where high quality and complex films, i.e. quantum wells, quantum wires, are required [20]. Achieving epitaxial film growth depends on a number of factors, such as equilibrium thermodynamics of nucleation, creation of vapor from reactants, substrate surface reactions, and mobility of species through the substrate surface. The complex interaction of these atomic processes may lead to the formation of epitaxial films only if they occur over a certain temperature called "epitaxial temperature", which depends on the specific system and deposition parameters. Epitaxial grow usually takes place at high substrate temperature to promote the mobility of adatoms in the substrate surface forming islands that become stable after they reach a certain size, and their continuous growth increases the nucleation density giving rise to a solid film with a preferential orientation. The quality of epitaxial growth is strongly influenced by the strain created in the film during deposition due to lattice misfit, and thermal strain produced as a result of the different coefficients of thermal expansion between the films and substrate. Substrate surface contamination also influences the growth by interrupting the formation of epitaxial layers. To prevent this, deposition is normally carried out under reduced pressures or vacuum to promote the effusion of impurities from the substrate.

4. Thin film morphology

Establishing the correlation between deposition parameters with the resulting morphology of deposited thin films is very complex because of the interaction of a number of factors, which influence the nucleation and growth phases. However, despite the wide spectrum for variability, it is possible to find some typical morphological features that are common in a broad range of thin films.

Probably one of the most common morphological features found in many thin films grown by vapor-based techniques is a columnar structure whose growing direction is not necessarily perpendicular to the substrate. The way how the adsorbed atoms are integrated into the growing process determine the final morphology of thin films. When the atoms strike the substrate surface they may contribute to the creation of islands. These islands increase in size due to the feeding of atoms coming from the diffusion in the substrate surface and also from the vapor flux, forming complex arrangements of islands or compact islands. The flux of atoms diffusing in the substrate is suppressed once the islands start an interfering process between each other due to the progressive growth, and only the contribution from the vapor flux remains active shifting the growth to the thickness direction. The accumulating growth in the thickness direction gives rise to a columnar structure with an appearance of cauliflower-like in the top view [4, 17, 21].

It was reported that, for the formation of a columnar structure with a rough surface, both a limited surface mobility of adatoms and a vapor flux arriving non-normal to the substrate are necessary conditions [4]. The former tends to control de size of the crystallites growing at the interface and defines the area at the base of the columns. The latter gives rise to a geometrical shadowing growth process. The atoms from the source of various deposition methods, i.e. CVD, sputtering, not always strike the substrate surface perpendicularly. This deviation from a normal incidence forces the development of a columnar growth in a direction towards the vapor flux. Overall, it has been observed that the angle between the substrate normal and the column direction falls in between the angle formed by the substrate normal and the vapor flux.

The evolution of the morphology in thin films as a function of substrate temperature (adatom mobility) can be classified according to a model introduced by Movchan and Demchishin in 1969 [22], named structure zone model (SZM). This model provides three structural zones depending on a homologous temperature (T_h) which results from the ratio between the substrate temperature (T_s) and the melting point temperature of the deposited material. In the zone I for a $T_h \leq 0.3$, the relative low substrate temperature allows for a low diffusion of adatoms which are not enough to compensate the density of defects. As a result, amorphous or nano-crystalline columns with high density of defects between them are formed with a cauliflower-like appearance in the top surface, which is enhanced by the shadowing process. In the zone II for a $0.3 < T_h < 0.5$, with a higher substrate temperature, the diffusion of adatoms is higher and enough to compensate the defects, producing a columnar structure with a lower concentration of defects and larger crystallites. In the zone III where $T_h > 0.5$, the higher substrate temperature produces a dominant bulk diffusion in the layer giving rise to a structure formed by coarse crystallites [23].

5. Properties of thin films

In this section a brief description of the mechanical, electrical, and optical properties in connection with the morphological features will be provided.

5.1 Mechanical properties

Thin films, due to their versatility to provide tailored properties, have found application in a number of sectors going from simple coatings for wear and corrosion protection, to more advanced applications such as antireflective coatings, microelectronics, photovoltaics, etc. Although these material structures have been selected due to exclusively their functional properties, they must be able to provide a reliable service operation with a proper mechanical and chemical resistance during the lifetime. These films, during deposition and operation, are prone to develop large stresses that might cause deformation and eventually mechanical failure, and therefore, it is essential to understand the microstructural processes involved in such effects to attempt to mitigate through the control of microstructure during the fabrication. Unlike bulk structures where the mechanical properties do not show a clear dependence on the sample size, in reduced structures like these, these properties are strongly affected by the resulting microstructure giving rise to a different behavior as compared to their bulk counterparts.

Let us assume a thin film deposited on relative tick substrate as illustrated in **Figure 1**. If strain by any means were produced on the film, it would change its dimensions relative to the substrate where it is deposited to maintain the equilibrium. If, hypothetically, the film were not adhered to the substrate, it would be visible the change in dimensions, for example, when the strain has expanded the original dimensions of the film, as shown in part b of **Figure 1**. The action of matching again the expanded film into the substrate entails the application of a deformation to force the film to adopt the substrate dimensions as shown in part c, giving rise to the generation of stress within the film. The stored stress naturally tends to be released to reach equilibrium, but depending on the degree of the substrate stiffness, part of this strain can be absorbed if the substrate is compliant which is reflected in a bending produced in the film/substrate system, or can remain entirely in the film if the substrate has a high stiffness.

The means by which strain can be produced in the films during deposition basically derive from the deposition method employed and from thermally-induced effects. Various vapor-based techniques such as sputtering and PECVD are well-known for producing stress due to the incorporation of gas into the microstructure or due to ion bombardment, whose degree can be controlled through a strategic balance of deposition parameters [24]. Impurity incorporation into the microstructure of a host material has also been ascribed as responsible for stress creation. Depending on their size, the introduction of dopant atoms cause a local deformation of the lattice producing stress. A notable example of this case is the fabrication of SiGe thin films [25], where the introduction of Ge atoms into the Si matrix increases the stress, which collaterally contributes to enhance the thermoelectric properties [26]. Thin film deposition is commonly realized at relative high temperatures and cooled down latter on. This change of temperature along with a difference in the coefficients of thermal expansion between the film and substrate material generate stress in the film microstructure as well [27]. For some thin films like hydrogenated microcrystalline silicon, widely used in the photovoltaic sector, large stress creation has been reported in the phase transition from amorphous to microcrystalline silicon [28].

The stress contained in the films affect the mechanical properties such as yield strength and hardness. The yield strength of thin films is reported to be higher than the bulk version due to the influence of the microstructure. The value of this property is reported to increase with a smaller grain size, and with a higher density of dislocations present in the microstructure [29]. The deformation mechanism model that may explain the strengthening of thin films is based on the dislocation motion.

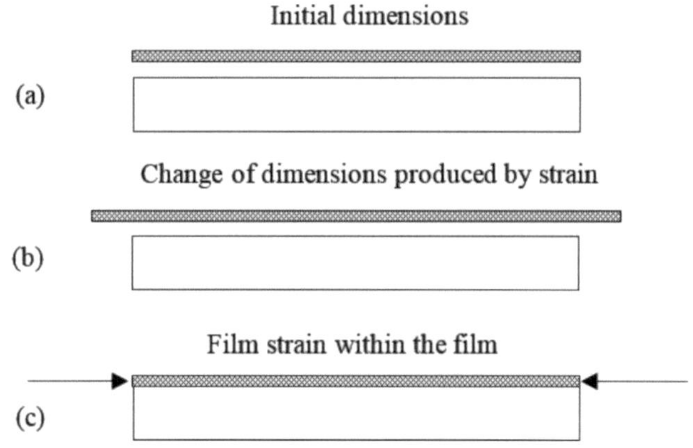

Figure 1.
Schematic of stress/strain creation within thin films.

A number of dislocations present in the microstructure move as a function of the stress applied to the film, and for a dislocation to move, the stress applied must be comparable or higher than the energy necessary to deposit a misfit dislocation. However, it is important to consider that dislocation motion can be constrained by the interaction with other microstructural defects such as extended defect, point defects, and other dislocations which contribute to film strengthening. When the film/substrate system is considered instead of free-standing films, the constraints to the mobility of dislocations provided by the substrate and any oxide present in the substrate surface have to be considered to determine the strength. These additional constraints have demonstrated to strengthen the film in comparison to free-standing ones [30]. These microstructural features also explain the larger hardness exhibited in thin films in comparison to bulk materials.

5.2 Electrical properties

Electrical properties within thin films comprise a broad field if one considers the different resulting microstructure, whether they are metallic films, semiconductors or insulators films, and the type of substrate on which they are deposited. However, much of these films possess some common morphological features that derive in similar transport mechanisms that allow to treat the conductivity in a global perspective.

One of the main factors for the deviation of conductivity in thin films with respect to bulk material is the size effect. The electron mean free path of the bulk material reduces as the material thickness reduces due to the activation of additional scattering mechanisms. In a simplified approximation, dictated by the direct proportionality of conductivity with the electron mean free path in bulk materials, the conductivity undergoes a reduction. This correlation works for either epitaxial, polycrystalline or amorphous structures as the maximum crystallite size is limited by the film thickness, preserving the size effects, considering that all the other constituent components of the conductivity remain unchanged [4, 31]. However, as described in previous sections, the microstructure might contain a large number of structural defects and grain boundaries which can act as scatters for charge carriers, further reducing the conductivity.

In one of the most complex microstructures where the film is a semiconductor formed by small crystallites embedded in amorphous tissue and surrounded by a large number of grain boundaries, and containing either electrons or holes as majority carries, the transport mechanism becomes very complicated owing to the simultaneous interaction of various scattering mechanisms. In this type of microstructure, a larger crystalline volume fraction favors the conductivity by allowing a higher carrier mobility unlike the amorphous phase. The grain boundaries are considered as a disordered region were mobile carriers are scattered in their travel between crystallites. As an example, the electrical conductivity of polycrystalline and nano-crystalline materials is substantially lower than the bulk single crystalline counterpart attributed to a reduced carrier mobility in spite of having a similar carrier concentration [32]. Microstructural defects such as voids, dangling bonds, and localized defects, which are found mostly in the grain boundaries, also contribute to further reduction of conductivity. These defects are known to trap mobile carriers of doped semiconductors, forming a potential energy barrier which limits the motion of charge carriers between crystallites [33]. Accordingly, electrical conductivity is reduced by the decrease in the number of free carriers available for conduction and a reduction of the carrier mobility. Additional scattering mechanisms appear in doped semiconductors such as ionized impurity scattering, carrier-carrier scattering at room temperature, and carrier-phonon scattering at high temperatures which further contribute to the reduction of electrical conductivity.

5.3 Optical properties

The optical behavior of thin films is determined by the resulting microstructure that depends on the deposition parameters. Before explaining the relationship of the optical properties with microstructural features, it is necessary to define the optical coefficients. The optical response of thin films can be characterized based on the reflection and transmission coefficients. In a general arrangement a thin film with a thickness d_1 and refractive index n_1 is deposited onto a substrate of similar/different material with a refractive index n_2. When a beam of light strikes the interfaces formed by the incident medium and the film, and the film-substrate interface, multiple reflections and transmissions occur at both interfaces at specific angles whose total amplitudes are computed from the sum of individual reflections directed back into the incident medium, and the individual refractions traversing into the substrate, respectively. The main difference between nonabsorbing and absorbing layers is that the refractive index from the nonabsorbing layer becomes more complex by integrating a quantity k named *extinction coefficient* in the absorbing layer, which defines the absorption of energy within the film. Thus, it can be inferred that the optical coefficients strongly depend on the refractive indexes of each medium, the extinction coefficient, and the film thickness.

Due to the consistent results reported in literature, it is possible to generalize the behavior of optical properties with film thickness even though specific details related to film deposition are not provided. Both refractive index and extinction coefficients show a strong dependence on the film thickness. For most metallic films, i.e. Au, Ag, while the former decreases from a value higher than that corresponding to a bulk material as the film thickness increases, the extinction coefficient tends to increase, from a very low value, approaching that of the bulk [4]. These optical coefficients also behave differently depending on the wavelength of the incident beam because they become dominated either for intra-band transitions in the visible, or free electrons in the infrared region. The electronic contribution to the optical behavior explains the effect of film thickness on the optical response. As the thickness is reduced the surface scattering mechanism increases, reducing in turn the electronic mean free path and the scattering time, which limits the contribution to the optical conductivity. The reduction of these both parameters produces an increase of electrical resistivity, and thus, a connection between optical properties and resistivity can be established. This correlation implies that structural defects such as voids, unsaturated bonds, point defects, extended defects, band tail states, and structural features like grain boundaries and oxide, all of which contribute to increase the carrier scattering, influence the optical properties.

The behavior of reflection and transmission is also a function of the degree of roughness in the film created during deposition [34, 35]. Essentially the roughness degrades the film uniformity producing a thickness variation across the film which affect the transmission and optical coefficients.

6. Characterization techniques

In order to understand the connection between deposition parameters, the resulting morphology and the physical properties of thin films, it is necessary to characterize each of these features. The reduced volume of thin films, however, does not allow to employ universally all the characterization techniques developed for bulk materials. Accordingly, in this section the most common characterization techniques suitable for thin film measurements are briefly described.

6.1 Mechanical characterization

The mechanical characterization comprises the determination of residual stress commonly created during deposition, and the physical properties such as Young's modulus and hardness.

The well-known Stoney Equation [36] is widely used to characterize the stress generated in a film deposited onto a thicker substrate. The principle behind this equation is the linear correlation between the stress crated within the film, and the amount of bending produced in the substrate as a result of the constraint condition among them. The application of this technique requires the knowledge of the radius of curvature of the substrate before and after the film deposition. The radius of curvature is often measure with an optical profilometer or drawing on any high resolution microscopy method that allows to reproduce the sample profile. The accuracy of this method is subjected to the compliance of certain conditions, i.e. the film thickness is much thinner than the substrate, very small strain and rotations, both film and substrate should be homogeneous and isotropic, equi-biaxial stress in the plane of the film, spherical deformation of the system film/substrate, and spatial invariability of stress and curvature in the whole surface. Even though some of these conditions are not fully in compliance, this technique has still been used for stress determination in a number of investigations including semiconductor-based thin films in the solar sector, thin film transistor industry among others [37, 38].

Physical properties such as hardness and Young's modulus are usually characterized drawing on the nano-indentation technique [29, 39]. This method consists in moving a sharp indenter towards the surface of the film until making an effective contact, then the applied load and the displacement are repeatedly recorded. In this way a correlation of the applied load through the indenter and the depth of indentation is established. As the indenter penetrates into the film, the slope of the loading curve progressively increases due to stronger contact between them; this correlation allows for the calculation of the hardness at any point of the curve by dividing the load to the contact area in that point. The Young's modulus of the film instead can be determined from the unloading curve once the contact area is defined considering that the Young's modulus and Poisson's ratio of the indenter are known. This is possible thanks to the direct correlation that is established between the backward motion of the indenter and the elastic properties of the film during the unloading.

6.2 Chemical composition

A number of techniques are available to obtain the elemental composition of thin films, which can be classified in two groups, ion scattering and spectroscopic-based techniques. In the first group both Rutherford backscattering spectrometry (RBS) and elastic recoil detection analysis (ERDA) are included. They both are based on the elastic scattering of energetic ions produced when they strike the atoms in the film surface; in this process they transfer an amount of energy to the target species atoms via collisions generating backscattered, forward scattered, and recoils particles. While in RBS the backscattered yield and its energy distribution are measured by a detector, in ERDA the corresponding quantities for the coiled particles are recorded allowing the formation of an energy spectrum from which the compositional depth profile can be extracted. These both techniques can be used as complement of each other as the RBS is suitable for detection of heavy elements while ERDA provides a higher accuracy for light elements. The main disadvantage of RBS is a shallow depth into the film that can be probed while in ERDS complications to distinguish elements with similar masses arise.

In the spectroscopic-based techniques, the X-ray photoelectron spectroscopy (XPS) is one of the most widely employed for elemental composition characterization. The XPS is based on photoelectric effect through which electrons localized either in the core or valence band are emitted when a beam of X-rays, with an energy higher than the binding energy of the electrons, is irradiated onto the film surface. Then, these ejected electrons are driven first to an analyzer to measure their kinetic energy, and subsequently they arrive to a detector where the number of electrons is counted considering their kinetic energies. The information of this kinetic energy along with the known photon energy allow to compute the electron binding energy. A spectra is then formed that correlates the electron count vs. the calculated binding energy. The binding energy constitutes a signature for the identification of each constituent element of the film since each element possesses a unique value. On the other side, the intensity of the spectra instead reflects the concentration of the element. This technique is suitable for probing the sample in a depth of a few nanometers as it is limited by the interactions of emitted electrons with the atoms present in the film.

6.3 Microstructure and morphology

A number of methods are available for the microstructure characterization in terms of crystalline volume fraction, crystallite size, and crystallographic orientation.

The distinctive microstructure and material phase of thin films as a result of deposition conditions can be extracted from Raman microscopy, and X-ray diffraction (XRD). The Raman technique [40] is a non-destructive method based on the scattering of incident photons, coming generally from a laser beam, caused by the vibrational modes of molecules or atoms present in the film. The scattering of photons can be either elastic, or inelastic; within the latter, photons can be scattered with a frequency lower than incident photons (stokes) or higher than incident photons (anti-stokes). A Raman spectrum is formed considering the intensity and the frequency of the scattered light (inelastic) corresponding to the specific vibrational mode of the molecule in question. The material phase fraction in volumetric terms can be extracted from the Raman line-shape. The different peaks observed in the Raman spectrum designates different material phases, which can be extracted by isolating the integrated intensity of the required peak. This isolation is usually carried out by a deconvolution of the Raman spectra using specialized software that allows to perform a peak fitting analysis. The average crystallite size can also be extracted from the Raman spectrum by means of a correlation length model which is suitable for sizes larger than 5 nm. However, certain considerations are necessary to avoid the superposition of simultaneous effects such as stress and local heating due to a high laser power.

The X-ray diffraction technique [41] is based on the scattering principle where a monochromatic beam of X-rays is directed onto a specimen which contains a set of lattice planes. The incident X-rays are scattered in different angles by the lattice planes according to the Bragg's law, giving rise to constructive interference which build up the peaks of the diffraction pattern. The peak shape defines the material phase for the substance under study providing a broad peak for amorphous regions, and sharp peaks for crystalline ones. In this way, the fraction of crystallinity can be determined once the integrated intensities of each characteristic peak is obtained. The peak width also provides information about the crystallite size, which shows a broadening for small crystallites and shrinking for larger ones. The crystallite size is computed drawing on the well-known Scherrer's formula. In the case of crystalline orientation it can be inferred from the relative change of the peak height.

DOI: http://dx.doi.org/10.5772/intechopen.95527

The morphology of the film can be investigated via scanning electron microscopy (SEM) and transmission electron microscopy (TEM). SEM is one of the most widely used tools to characterize the morphology of thin films. This technique is based on the interplay of a beam of electrons with the sample surface. This interaction produces the emission of electrons from the sample with different energies due to either elastic, inelastic scattering or photons, which are collected by a detector to produce a distribution map based on the intensities of the signal. Each of the three emitted scattered electrons allows the reproduction of images which provide different information about the sample, i.e. images from secondary electrons (inelastic scattering) are suitable for the study of topographical features. From a top view this technique permits the visualization of the crystalline columns emerging at the sample surface, and the different material phases, either crystalline and/or amorphous, due to the contrast created by the secondary electrons, while the film thickness can be evaluated form a cross-sectional SEM analysis.

TEM consists in the emission of a beam of electrons which are directed onto a very thin sample (a few hundred nm) to allow the transmission of such electrons. The intensity with which these transmitted electrons exit the sample depends on the density and thickness of the sample; thus, these structure-dependent intensities give rise to the formation of a contrast that is projected as an image on a fluorescent screen. In essence, a compact structure produces a higher scattering of electrons projecting a darker image while a porous structure projects a brighter image in a bright field image dominated exclusively by transmitted electrons. TEM constitutes a powerful technique with a spatial resolution higher than SEM for the characterization of microstructure, crystallite size, crystalline orientation, and the film thickness can be obtained via a cross-sectional TEM image.

In some characterization cases two or more techniques are complementarily employed to extract thorough information from thin films. For example, due to the in-depth probing the Raman technique can provide the size of the small crystallites that form a columnar structure in the film while the SEM can be used to determine the diameter of the crystalline columns emerging at the sample surface. Furthermore, the selection of the appropriate characterization technique strongly depends on the characteristics of the film such as the thickness, whether it is a conductor, semiconductor or insulator, whether it is possible to achieve a high vacuum, among others.

6.4 Electrical properties

Perhaps the most widely used technique to measure the electrical resistivity is the four-point probe method. It consists of four metal tips linearly arranged keeping the same separation from each one where an electric current is applied in the outer two probes while the potential difference is measured in the inner two probes. From these measurements the sheet resistance (R_{sh}) of the film can be calculated by means of the following equation. The Sheet resistance is defined as the ratio of the resistivity (ρ) to the film thickness (t).

$$R_{sh} = \frac{\rho}{t} \tag{1}$$

An extra issue related to the measurement of carrier concentration and carrier mobility appears in semiconductor characterization. The Van der Paw technique [42] provides a solution to obtain these parameters. This method is based on two independent measurement, i.e. resistivity and Hall coefficient. The characteristic resistances are first obtained by the application of a current in two adjacent contacts while the potential difference is measured in the other two remaining contacts, all of which are located in the periphery of a sample with an arbitrary shape. Then these results are incorporated into an expression developed by Van der pauw where d is the film thickness, f is a correction factor that depends on the ration of the characteristic resistances R_A and R_B.

$$\rho = \frac{\pi d}{\ln 2}\left(\frac{R_A + R_B}{2}\right) f \tag{2}$$

The Hall effect consists in the creation of a voltage when an electric current (I) is applied between opposite contacts of a semiconductor that is under the effect of a magnetic field perpendicular to the plane of the sample (B). This potential difference is created by the migration of charge carriers, either electrons or holes, to the edge of the sample induced by the magnetic force. The sign of this voltage is determined by the type of major carriers dominating the electronic transport. This Hall voltage (V_H) is used to obtain the Hall coefficient (R_{HS}) and the carrier concentration (N_S) through the following equation.

$$R_{HS} = \frac{V_H}{IB} = \frac{r}{qN_S} \tag{3}$$

where r is the scattering factor, and q is the elementary charge.

The carrier Hall mobility (μ_H) is obtained by combining the Hall coefficient with the resistivity previously calculated

$$\mu_H = \frac{R_{HS}}{\rho} \tag{4}$$

7. Applications and challenges of thin films

Thin films technology has historically been used in a wide range applications going from decorative purposes in its early stage, evolving for optical purposes latter on, and an almost endless range of applications with the appearance of advanced deposition techniques, supported by the rapid development of vacuum technology and electrical power. Overall, thin films are used to enhance the properties of bulk materials by depositing a layer with the desired physical and chemical characteristics to improve their functionality. In the following section a brief description of the most technological relevant fields of application of thin films is presented.

Advanced electronics-optoelectronic devices have become an important field for the application of a number of thin film types. In particular, MOSFET and CMOS absorb a great amount of the technological development in semiconductor thin films. The fabrication of MOSFETs requires the use of dielectric thin films, i.e. silicon dioxide (SiO_2) [43, 44] to insulate the conducting channel from the gate. This thin film has been used due to its ease of fabrication, high impedance due to a large band gap, resistance to high temperatures and chemicals. Also, metallic films are required for the fabrication of multiple microelectronic devices, opto-electronics and optical devices [13, 45]. Al thin films are usually deposited in the channel between the source and drain of MOSFETs to allow the voltage for its operation. Instead, Cu thin films are commonly used in CMOS as gate metallization due to its high electrical conductivity and higher resistance to electromigration. Thin films have also played an important role in data storage devices due to their good magnetic properties in an attempt to replace the traditional flash memory devices for non-volatile memory devices [46]. Thin films such as $BiFeO_3$, lead-zirconium titanate films, amorphous Si, organic compounds among others are being explored as candidates as based-material for this application [47–50].

The use of *thin films in the photovoltaic sector (PV)* is conceived as a potential solution to reduce the cost per watt in the generation of electricity. This sector has been experiencing a rapid market penetration due to the accelerated achievement of higher efficiencies and the development of thin film structures with better stability. In fact, record efficiencies about 23.3% and 22.1% have been reached using copper indium gallium selenide (CIGS) and CdTe thin films as based materials, respectively [51]. Overall, the advantages of thin films in the PV sector is related to the high absorption coefficient of the absorber layer, which permits to reduce considerably the material thickness, contributing to the reduction of material cost; also thin film technology allows the deposition of multiple-juntion devices to capture most of the solar spectrum to increase the conversion efficiency [52]. Additionally, thin films can be deposited into flexible substrate for roll-to-roll manufacturing of PV modules [53]. The state-of-the art of thin films for PV application was initially dominated by amorphous silicon, but evolved into the more efficient CdTe and CIGS, and lately organic and perovskite-based PV cells are under investigation due to its reduced processing cost and feasibility to deposit at low temperatures in flexible substrates [54, 55].

Thins films and coating applications are involved in a large number of fields including optics, and in sectors where the improvement of the mechanical and chemical properties of bulk materials provide a better functionality or larger lifespan. Optical thin films are widely employed in eyeglasses to improve the vision through the use of a polymer-based optical element that is coated to the spectacles. In addition, the undesired transmission of ultraviolet light and undesired reflection are prevented by the use coatings materials able to absorb wavelengths lower than 400 nm, and the use of antireflective coatings usually made of dielectric materials [56]. Architectural glazing has drawn on thin film coatings to enhance the energy efficiency in office buildings [57, 58]. The heat transfer can be managed from outside and inside the buildings by a suitable filtering of the spectral regions of light. Only transmission of visible light from the outside, and reflection of infrared radiation from inside can be set by making the windows to become a multifunctional device-like with thin films with different spectral response, saving energy from air conditioning and heating for the former and latter cases, respectively. Coatings as a means to increase the wear resistance and reduction of friction in cutting tools can be obtained by multilayer deposition of ceramic coatings, i.e. TiN, TiC [13]. Coatings for corrosion resistance are widely spread in numerous sectors including pipes coated with SiC, stainless steel components coated with oxides, i.e. SiO_2, Al_2O_3, engine parts coated with high temperature corrosion protection such as $MoSi_2$ among others [13] .

Organic thin films have attracted a great attention owing to certain unique properties, in particular, flexibility and low cost material processing which are essential to expand the scope of application of many technologies. In photovoltaics for example, although still low, the efficiency has been improved considerably from 0.04% up to about 8.3% [54] in organic-polymer based modules, but its evolution remains fuelled by the low cost of material processing, i.e. printing, spraying, and the possibility to fabricate flexible modules. Likewise, the intrinsic complex fabrication process and rigidity of Si-based field-effect transistors can be somehow overcome by organic thin films field-effect transistors. The use of organic thin film have already been proven in various applications such as memory devices, sensors, electronic papers, and smart cards [59, 60].

The use of thin films has gained a considerable space in *biomedical applications* due to their ability to provide biocompatible and functional properties, for example, invasive devices, tissue engineering substrates, drug delivery, and antimicrobial coatings, to name a few. The surface of implants have to comply special chemical and mechanical properties, and Ti6Al4V thin films appear to provide appropriate conditions for femur implants [61]. This structure apart from offering a good adhesion and harness, promotes the formation of a calcium layer through a chemical interaction with the biological fluids, improving the osseointegration. Polymer-based thin films have demonstrated to have a good resistance to protein adsorption, which is essential to provide a biocompatible behavior to implants. In this respect, poly (ethylene glycol) PEG, PEGylated thin films are suitable for bone, dental implants and for tissue engineering purposes [62, 63]. Composite thin films have also been used to provide the appropriate mechanical and biological properties to implants in neuronal applications. For example, silicon-based implants have been coated with a nanostructure formed by amorphous silica with fillers of aluminum, silicon dioxide or silver in order to provide microbial protection. Inorganic thin films with piezoelectric properties deposited on flexible substrates are also being investigated for the fabrication of nano-generators and nano-sensors for biomedical applications [64]. These piezoelectric devices have the capacity to convert mechanical energy provided by the movement of internal organs into electrical energy to power for example pacemakers or nano-sensors. Due to the high sensitivity to mechanical movement these devices can also be used to monitor the cell deformation at nanoscale. Higher performance piezoelectric devices have been fabricated using perovskite such as $BaTiO_3$, PZNT, and PMN-PT [64, 65].

The broad scope of thin film applications require of tailored physical, mechanical and chemical properties which are linked to the resulting structure and morphology, and they in turn depend on the deposition techniques and deposition parameters adopted. Accordingly, a number of challenges remain to be tackled for a complete understanding of the connection among the different phases involved in the fabrication of thin films. Overall, various versions of CVD and PVD deposition techniques present a still expensive final product, lack of reproducibility, inappropriate attachment of the film to the substrate, high deposition temperatures which prevent the use of cheaper substrates, and limited control over the final properties. Thus, deposition technology needs to evolve with a higher precision to control the microstructure, and with a higher deposition rate suitable for large area deposition to reduce the cost. Moreover, although nanostructured thin films are promising for cutting edge applications such as microelectronics, optics, photovoltaics, and biomedicine, some of them need to be transferred to specific substrates for an appropriate operation. Consequently, the now poor transferring technique has to progress to take fully advantage of thin film technology. Even though a number of characterization techniques are available for tracking almost every feature of thin films, many of them struggle when the film thickness approaches a few nanometers. For example Raman or X-ray diffraction characterization might degrade due to the inevitable contribution from the substrate to the acquired spectra. Sophisticated models to analyze the data are therefore needed to isolate the relevant information. Of primary importance for biomedical applications is the knowledge of protein adsorption in substrates for the appropriate selection of materials; however, the characterization tools for this purpose are still at their infant stage, and they are based on complicated models for data analysis. Therefore, more advanced and specific in vitro models can pave the way for a rapid identification of suitable thin films. In spite of the significant progress in deposition and characterization techniques, the prediction of film properties as a function on the microstructure is still very difficult. This occurs due to the complex transport properties derived from the multiple defects, grain boundaries, material phases, quantum confinement effects in very thin films, interface scattering, among others. Thus, advanced models that incorporate all of these structural variants are necessary to establish the appropriate connection between microstructure and film properties to progress in the thin film technology.

8. Conclusions

This chapter attempted to carry out the study of thin films maintaining a tight correlation among the deposition techniques, the resulting microstructure, and their physical properties. It was found that the development of deposition methods are driven to satisfy the needs for films with specific mechanical, physical and chemical properties. Overall, they have evolved to enable the fabrication of thin films with increasingly higher purity, based on a variety of depositing materials and substrates with high reproducibility. However, the cost of material processing of certain structures such as nanostructures, quantum well, quantum wires, and quantum dots is still a challenge for commercial applications. The selected deposition technique and deposition parameters greatly define the final thin film microstructure being either amorphous, polycrystalline or epitaxial. In particular, the deposition temperature and deposition rate strongly influence the adatom-diffusion at the substrate surface giving rise to the formation of islands whose sizes increase with feeding of atoms. The most common morphological feature in many films is a columnar structure formed due to the feeding of atoms from the vapour flux once the adatom-diffusion in the substrate surface has placated due to the interfering process between islands. The mechanical, electrical, and optical properties of thin films are determined by the final morphology. It is important to remark that these properties normally deviate from the values corresponding to their bulk counterparts. For example, the yield strength and hardness are higher in thin films due to the influence of crystallite size, and movement of dislocations usually found in thin films. Likewise, the electrical conductivity is also affected by additional scattering mechanisms appearing due to the reduced film thickness. There are several techniques available for thin film characterization, but their accuracy strongly depends on the depositing material, substrate material, and film thickness. In most cases, two or more techniques are complementary used to access the required information with high precision. Regardless of the progress in thin film technology, important challenges remain to be tackled, including the accurate prediction of film properties based on the final microstructure, more advanced characterization techniques in the biomedicine field, and sophisticated models for data analysis.

Acknowledgements

This chapter was produced with the financial support of SENESCYT-ECUADOR, with contract number 564-2012.

Author details

Edwin Acosta
Heriot Watt University, SENESCYT-ECUADOR, Edinburgh, UK

*Address all correspondence to: epa30@hw.ac.uk; edwin_paf@hotmail.com

References

[1] Hartmut Frey and Hamid R. Khan: *Handbook of Thin Film Technology*. (Berlin, Heidelberg: Springer Berlin/ Heidelberg, Berlin, Heidelberg, 2015).

[2] Hans K. Pulker, (Berlin, Heidelberg: Springer Berlin Heidelberg: Berlin, Heidelberg), pp 353-369.

[3] John R. Arthur, In *Specimen Handling, Preparation, and Treatments in Surface Characterization*, ed. Alvin W. Czanderna, Powell Cedric J., Madey Theodore E., Hercules David M. and Yates John T. (Springer US: Boston, MA, 1998), pp. 239-293.

[4] Robert D. Gould, Safa Kasap and Asim K. Ray, In *Springer Handbook of Electronic and Photonic Materials*, ed. Safa Kasap and Capper Peter (Springer International Publishing: Cham, 2017), pp 1-1.

[5] S. Ghamaty, J. C. Bass and N. B. Elsner, In Proceedings ICT'03. 22nd International Conference on Thermoelectrics (IEEE Cat. No.03TH8726), (2003), pp. 563-566.

[6] Allon I. Hochbaum, Renkun Chen, Raul Diaz Delgado, Wenjie Liang, Erik C. Garnett, Mark Najarian, Arun Majumdar and Peidong Yang, Enhanced thermoelectric performance of rough silicon nanowires, Nature, vol. 451, 163-167 (2008)

[7] Akram I. Boukai, Yuri Bunimovich, Jamil Tahir-Kheli, Jen-Kan Yu, William A. Goddard Iii and James R. Heath, Silicon nanowires as efficient thermoelectric materials, Nature, vol. 451, 168-171 (2008)

[8] W. Gulbiński, In *Chemical Physics of Thin Film Deposition Processes for Micro- and Nano-Technologies*, ed. Yves Pauleau (Springer Netherlands: Dordrecht, 2002), pp. 309-333.

[9] GS Anderson, Wm N Mayer and GK Wehner, Sputtering of dielectrics by high-frequency fields, Journal of Applied Physics, vol. 33, 2991-2992 (1962)

[10] SA Awan and RD Gould, Conductivity and dielectric properties of silicon nitride thin films prepared by RF magnetron sputtering using nitrogen gas, Thin Solid Films, vol. 423, 267-272 (2003)

[11] Y Ochiai, K Aso, M Hayakawa, H Matsuda, K Hayashi, W Ishikawa and Y Iwasaki, On specimen preparation by radio frequency planar magnetron sputtering, Journal of Vacuum Science & Technology A: Vacuum, Surfaces, and Films, vol. 4, 19-25 (1986)

[12] Michael NR Ashfold, Frederik Claeyssens, Gareth M Fuge and Simon J Henley, Pulsed laser ablation and deposition of thin films, Chemical Society Reviews, vol. 33, 23-31 (2004)

[13] K. L. Choy, Chemical vapour deposition of coatings, Progress in Materials Science, vol. 48, 57-170 (2003)

[14] In *Encyclopedia of Nanotechnology*, ed. Bharat Bhushan (Springer Netherlands: Dordrecht, 2012), pp. 146-146.

[15] N. Sharma, M. Hooda and S. K. Sharma, Synthesis and Characterization of LPCVD Polysilicon and Silicon Nitride Thin Films for MEMS Applications, Journal of Materials, vol. 2014, p. 954618 (2014)

[16] Milton Ohring: *The materials science of thin films / Milton Ohring*. (Academic Press Harcourt Brace Jovanovich Publ, San Diego New York Boston (Mass.) [etc, 1992).

[17] Matthew Pelliccione: *Evolution of Thin Film Morphology Modeling and Simulations / by Matthew Pelliccione,*

Toh-Ming Lu. 1st ed. 2008. ed. (New York, NY : Springer New York : Imprint: Springer, New York, NY, 2008).

[18] K Valalaki, N Vouroutzis and AG Nassiopoulou, Significant enhancement of the thermoelectric figure of merit of polycrystalline Si films by reducing grain size, Journal of Physics D: Applied Physics, vol. 49, p. 315104 (2016)

[19] Jun Ma and Sanjiv Sinha, Thermoelectric properties of highly doped n-type polysilicon inverse opals, Journal of Applied Physics, vol. 112, p. 073719 (2012)

[20] M. S Dresselhaus, G. Chen, M. Y Tang, R. G Yang, H. Lee, D. Z Wang, Z. F Ren, J. P. Fleurial and P. Gogna, New Directions for Low-Dimensional Thermoelectric Materials, Advanced Materials, vol. 19, 1043-1053 (2007)

[21] R. Messier and S. Trolier-McKinstry, In *Encyclopedia of Materials: Science and Technology*, ed. K. H. Jürgen Buschow, Cahn Robert W., Flemings Merton C., Ilschner Bernhard, Kramer Edward J., Mahajan Subhash and Veyssière Patrick (Elsevier: Oxford, 2001), pp. 9306-9313.

[22] Anna Kameneva, Evolution of the Film Structure in the Various Evaporation Processes, *Research Journal of Pharmaceutical,* Biological and Chemical Sciences, vol. 6, 951-960 (2015)

[23] Peter A. Dearnley and Tomasz Liskiewicz, In *Encyclopedia of Tribology*, ed. Q. Jane Wang and Chung Yip-Wah (Springer US: Boston, MA, 2013), pp. 3948-3960.

[24] Ciprian Iliescu, Marioara Avram, B. Chen, Alina Popescu, V. Dumitrescu, Daniel Poenar, A. Sterian, Danilo Vrtacnik, S. Amon and P. Sterian, Residual stress in thin films PECVD depositions: A review, Journal of Optoelectronics and Advanced Materials, vol. 13, 387-394 (2011)

[25] FC Marques, P Wickboldt, D Pang, JH Chen and W Paul, Stress and thermomechanical properties of amorphous hydrogenated germanium thin films deposited by glow discharge, Journal of applied physics, vol. 84, 3118-3124 (1998)

[26] Davide Beretta, Neophytos Neophytou, James M. Hodges, Mercouri G. Kanatzidis, Dario Narducci, Marisol Martin- Gonzalez, Matt Beekman, Benjamin Balke, Giacomo Cerretti, Wolfgang Tremel, Alexandra Zevalkink, Anna I. Hofmann, Christian Müller, Bernhard Dörling, Mariano Campoy-Quiles and Mario Caironi, Thermoelectrics: From history, a window to the future, Materials Science and Engineering: R: Reports, vol. 138, p. 100501 (2019)

[27] John A. Thornton and D. W. Hoffman, Stress-related effects in thin films, Thin Solid Films, vol. 171, 5-31 (1989)

[28] Hiroyuki Fujiwara, Michio Kondo and Akihisa Matsuda, Nucleation mechanism of microcrystalline silicon from the amorphous phase, Journal of non-crystalline solids, vol. 338, 97-101 (2004)

[29] William D. Nix, Mechanical properties of thin films, Metallurgical Transactions A, vol. 20, p. 2217 (1989)

[30] W. D. Nix, (2014), pp 353-421.

[31] Safa Kasap and Peter Capper: *Springer Handbook of Electronic and Photonic Materials*. (Cham: Springer International Publishing AG, Cham, 2017).

[32] N. D. Arora, J. R. Hauser and D. J. Roulston, Electron and hole mobilities in silicon as a function of concentration and temperature, IEEE Transactions

on Electron Devices, vol. 29, 292-295 (1982)

[33] John Y. W. Seto, The electrical properties of polycrystalline silicon films, Journal of Applied Physics, vol. 46, 5247-5254 (1975)

[34] Guowen Ding, César Clavero, Daniel Schweigert and Minh Le, Thickness and microstructure effects in the optical and electrical properties of silver thin films, *AIP Advances*, vol. 5, p. 117234 (2015)

[35] A. Ignatiev, P. O'Neill and G. Zajac, The surface microstructure optical properties relationship in solar absorbers: black chrome, Solar Energy Materials, vol. 1, 69-79 (1979)

[36] George Gerald Stoney and Charles Algernon Parsons, The tension of metallic films deposited by electrolysis, Proceedings of the Royal Society of London. Series A, Containing Papers of a Mathematical and Physical Character, vol. 82, 172-175 (1909)

[37] R. B. Wehrspohn, S. C. Deane, I. D. French, I. Gale, J. Hewett, M. J. Powell and J. Robertson, Relative importance of the Si–Si bond and Si–H bond for the stability of amorphous silicon thin film transistors, Journal of Applied Physics, vol. 87, 144-154 (1999)

[38] K. Christova, Sashka Alexandrova, A. Abramov, E. Valcheva, Bogdan Ranguelov, C. Longeaud, S. Reynolds and Pere Cabarrocas, Structure-related strain and stress in thin hydrogenated microcrystalline silicon films, Journal of Physics: Conference Series, vol. 253, p. 012056 (2010)

[39] P. S. Alexopoulos and T. C. O'Sullivan, Mechanical Properties of Thin Films, Annual Review of Materials Science, vol. 20, 391-420 (1990)

[40] J. C. Tsang, In *Dynamics on Surfaces*, ed. Bernard Pullman, Jortner Joshua, Nitzan Abraham and Gerber Benjamin (Springer Netherlands: Dordrecht, 1984), pp. 379-393.

[41] Armin Segmüller, I. C. Noyan and V. S. Speriosu, X-ray diffraction studies of thin films and multilayer structures, Progress in Crystal Growth and Characterization, vol. 18, 21-66 (1989)

[42] L. J. van der Pauw, In *Semiconductor Devices: Pioneering Papers*, (WORLD SCIENTIFIC: 1991), pp. 174-182.

[43] Fu-Chien Chiu, Tung-Ming Pan, Tapas Kumar Kundu and Chun-Hsing Shih, Thin Film Applications in Advanced Electron Devices, *Advances in Materials Science and Engineering*, vol. 2014, p. 927358 (2014)

[44] M. C. Rao and M. S. Shekhawat, A BRIEF SURVEY ON BASIC PROPERTIES OF THIN FILMS FOR DEVICE APPLICATION, International Journal of Modern Physics: Conference Series, vol. 22, 576-582 (2013)

[45] T. Usui, H. Nasu, S. Takahashi, N. Shimizu, T. Nishikawa, M. Yoshimaru, H. Shibata, M. Wada and J. Koike, Highly reliable copper dual-damascene interconnects with self-formed MnSi/ sub x/O/sub y/ barrier Layer, IEEE Transactions on Electron Devices, vol. 53, 2492-2499 (2006)

[46] N. Setter, D. Damjanovic, L. Eng, G. Fox, S. Gevorgian, S. Hong, A. Kingon, H. Kohlstedt, N. Y. Park, G. B. Stephenson, I. Stolitchnov, A. K. Taganstev, D. V. Taylor, T. Yamada and S. Streiffer, Ferroelectric thin films: Review of materials, properties, and applications, Journal of Applied Physics, vol. 100, p. 051606 (2006)

[47] Gustau Catalan and James F Scott, Physics and applications of bismuth ferrite, Advanced materials, vol. 21, 2463-2485 (2009)

[48] Pi-chun Juan, Yu-ping Hu, Fu-chien Chiu and Joseph Ya-min Lee, The charge trapping effect of metal-ferroelectric

(PbZr0.53Ti0.47O3)-insulator (HfO2)-silicon capacitors, Journal of Applied Physics, vol. 98, p. 044103 (2005)

[49] Fu-Chien Chiu, Wen-Chieh Shih and Jun-Jea Feng, Conduction mechanism of resistive switching films in MgO memory devices, Journal of Applied Physics, vol. 111 (2012)

[50] Fu-Chien Chiu, Peng-Wei Li and Wen-Yuan Chang, Reliability characteristics and conduction mechanisms in resistive switching memory devices using ZnO thin films, Nanoscale Research Letters, vol. 7, p. 178 (2012)

[51] Senthilarasu Sundaram, Katie Shanks and Hari Upadhyaya, In *A Comprehensive Guide to Solar Energy Systems*, (Elsevier: 2018), pp. 361-370.

[52] J. Hemalatha, A Review of: "The Physics of Solar Cells, Jenny Nelson", Materials and Manufacturing Processes, vol. 23, 735-736 (2008)

[53] G Khrypunov, A Romeo, F Kurdesau, DL Bätzner, H Zogg and Ayodhya N Tiwari, Recent developments in evaporated CdTe solar cells, Solar energy materials and solar cells, vol. 90, 664-677 (2006)

[54] Frederik C. Krebs, Fabrication and processing of polymer solar cells: A review of printing and coating techniques, Solar Energy Materials and Solar Cells, vol. 93, 394-412 (2009)

[55] Mulmudi Hemant Kumar, Natalia Yantara, Sabba Dharani, Michael Graetzel, Subodh Mhaisalkar, Pablo P Boix and Nripan Mathews, Flexible, low-temperature, solution processed ZnO-based perovskite solid state solar cells, Chemical Communications, vol. 49, 11089-11091 (2013)

[56] J. Larruquert, In *Optical Thin Films and Coatings*, ed. Angela Piegari and Flory François (Woodhead Publishing: 2013), pp 290-356.

[57] Mehran Arbab and James J. Finley, Glass in Architecture, International Journal of Applied Glass Science, vol. 1, 118-129 (2010)

[58] T. W. Boentoro and B. Szyszka, In *Optical Thin Films and Coatings*, ed. Angela Piegari and Flory François (Woodhead Publishing: 2013), pp. 540-563.

[59] B. Chandar Shekar, Jiyeon Lee and Shi-Woo Rhee, Organic thin film transistors: Materials, processes and devices, Korean Journal of Chemical Engineering, vol. 21, 267-285 (2004)

[60] ZhiGang Li, XinYan Zhao, Xin Lu, ZhiQiang Gao, BaoXiu Mi and Wei Huang, Organic thin-film solar cells: Devices and materials, Science China Chemistry, vol. 55, 553-578 (2012)

[61] Ana P Piedade, Francisco Romeu, Rita Branco and Paula V Morais, In *Thin Film Synthesis and Application for Medical and Biological Use*, (IntechOpen: 2018).

[62] Venkat K Vendra, Lin Wu and Sitaraman Krishnan, Polymer thin films for biomedical applications, Nanomaterials for the life sciences, vol. 5 (2010)

[63] Yasser Zare and Iman Shabani, Polymer/metal nanocomposites for biomedical applications, Materials Science and Engineering: C, vol. 60, 195-203 (2016)

[64] Geon-Tae Hwang, Myunghwan Byun, Chang Kyu Jeong and Keon Jae Lee, Flexible piezoelectric thin-film energy harvesters and nanosensors for biomedical applications, Advanced healthcare materials, vol. 4, 646-658 (2015)

[65] Jue Gong and Tao Xu, In *Revolution of Perovskite: Synthesis, Properties and Applications*, ed. Narayanasamy Sabari Arul and Nithya Vellalapalayam Devaraj (Springer Singapore: Singapore, 2020), pp. 95-116.

Section 2

Characterization

Chapter 2

Infrared Characterization and Electrochemical Study of Silanes Grafted into Surface of Copper

Mohamed Masmoudi

Abstract

The formation of a protective layer of tow silane coupling agents: γ -methacryloxypropyltrimethoxysilane (γ-MPS) or γ-aminopropyltriethoxysilane (γ-APS) on copper is studied by diffuse reflectance infrared spectroscopy (DRIFT), electrochemical (Potentiodynamic polarization) and gravimetric chemical (Weight loss) measurements. Dried in ambient conditions, the silane adsorbed on the copper subtract physically, however its protective action is not reliable. Thiolate and siloxane band formation ameliorate the protective action of the silanic layer on the surface of copper especially after curing process. Potentiodynamic polarization and Weight loss experiments show that the performances protective action of cured treatment (cured/Cu-silane) is higher than that of aging process (aged/Cu-silane).

Keywords: hybrid coating, copper, corrosion, silanic compounds, infrared spectra

1. Introduction

The corrosion behavior of copper and its alloys has been extensively studied for a wide range of experimental conditions. For example, there are the dominant materials for sea water systems in many countries. In chloride-rich media, the naturally oxide formed layer is insufficient to improve the anticorrosion propriety and coating is required for providing a higher protection level. Surface treatment based on the use of chromate conversion layer is efficient for corrosive protection of metal [1]. However, toxic and carcinogenic characteristics of chromate and similar hexavalent chromium compounds resulted in legally limited use.

Organic coating adhesion to metal substrates is one of the most important physical properties for corrosion resistance. Recently, the corrosion protective performance of coupling agent attracts the attention of several researchers. In fact, silane surface treatment is can be considered as an environmentally friendly alternative to chromatation [2].

Silane coupling agents are commonly used for the pre-treatment of metal surfaces to increase adhesion between polymeric (i.e., the paint) and inorganic materials (i.e., the metal). This twofold function of the coupling agent is due to their chemical structure, corresponding to general formula $R–Si–(OR')_3$, where R is an organo-functional group (amine, mercapto, epoxy, methacrylic, vinyl, etc.) whose role is to establish a chemical band with the organic coating and R' stands for an alkyl group (generally Et or Me) [3].

The principle of the adhesion of silane coupling agents on inorganic specimens is based on the siloxane function (Si–OR') reactivity. Th later is hydrolysed to Si–OH. Two reactions (hydrolysis: Eq. (1) and condensation: Eqs. (2) and (3)) take place [4]:

$$RSi(OR')_3 + 3H_2O \rightarrow RSi(OH)_3 + 3R'OH \qquad (1)$$

$$RSi(OH)_3 + (OH)_3SiR \rightarrow RSi(OH)_2\text{–}O\text{–}(OH)_2SiR + H_2O \qquad (2)$$

$$RSi(OH)_3 + (R'O)(OH)_2SiR \rightarrow RSi(OH)_2\text{–}O\text{–}(OH)_2SiR + R'OH \qquad (3)$$

These three reactions display the importance of hydrated environment in the condensation reactions taking place between the coupling agent and the metallic surface, and which will take place on the copper surface according to the following reactions (Eqs. (4) and (5)):

$$RSi\text{–}OH + OH\text{–}Cu(surface) \rightarrow RSi\text{–}O\text{–}Cu + H_2O \qquad (4)$$

$$RSi\text{–}OR' + OH\text{–}Cu(surface) \rightarrow RSi\text{–}O\text{–}Cu + R'OH \qquad (5)$$

The object of the present study is to investigate the adsorption of γ-MPS and γ-APS onto pure copper at different dried conditions and to study the possibility of using these two-coupling agent's treatment on copper in order to enhance its corrosion protective performance.

2. Experimental

For all experiments, we made use of copper sample of purity 99.99%. The copper electrodes were masked by epoxy resin, leaving 28.3 mm^2 as the working surface. Before each test, the exposed surface of copper was polished by SiC paper to # 1200 and rinsed with distilled water, degreased with acetone, hot air flux dried and instantly plunged in the silanic solution. The objective of all these cleaning steps is to remove of contamination and to introduce the hydroxyl group into a copper surface. The silane coupling agent solution was obtained by dissolving γ-MPS or γ-APS (5% vol./vol.) in ethanol (90% vol./vol.) and distilled water (5% vol./vol.) at ambient temperature for 10 min.

Prior to coating treatment, the coupling agent solutions were stirred for 120 min to allow the hydrolysis of the Si-OC_2H_5 groups. The organosilanes employed in this work wereγ -methacryloxypropyltrimethoxysilane, (γ-MPS) andγ -aminopropyl-triethoxysilane, (γ-APS), with a high purity (98%). The chemical structures of these compounds are schematically represented in **Figure 1**.

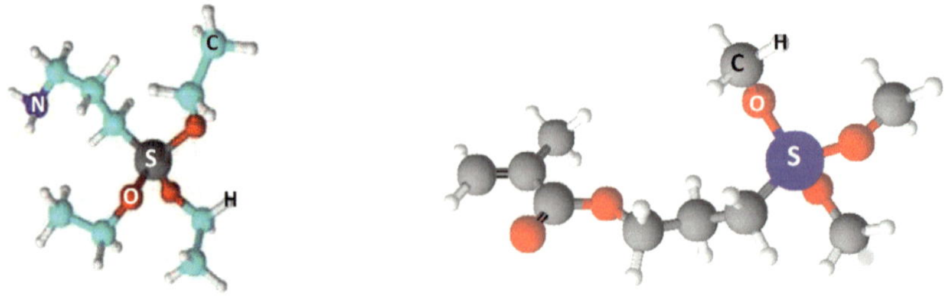

Figure 1.
The molecular structure of γ-MPS (a) and γ-APS (b).

All experiments were executed only on dipped substrate and dried at room temperature for 1440 min (aged/Cu-silane) and cured at 100°C for 60 min (cured/Cu-silane). After silanization step, all specimens were washed with ethanol for 5 min.

The Fourier transformed infrared (FT-IR) analysis was carried out using a Perkin-Elmer Paragon 1000 FT-IR spectrometer. Therefore, in order to analyze the chemical groups presented on the surface, we used the diffuse reflectance mode (DRIFT). The obtained spectra of the modified substrate were obtained by means of the diffuse reflectance accessory. The spectrum was recorded with a resolution of 2 cm^{-1}, with a total of 20 scans. The spectra were treated using BOMEM GRAMS software for the deconvolution of different region of the spectrum.

Potentiodynamic polarization curves were plotted from −400 to 600 mV/SCE at a polarization scan rate of 0.5 mV/second. Before all experiments, the potential was stabilized at free potential during 30 min. To guarantee the reproducibility, at least four replicates were run for each experiment. Electrochemical measurements were taken using an electronic potentiostat galvanostat radiometer controlled using the Volta Lab software. While, the electrochemical analyses were conducted under computer control.

For all tests, electrochemical experiments were performed using a standard three-electrode cell configuration. A saturated calomel electrode (SCE) was used as the reference electrode and a platinum sheet with 2 cm^2 totals surface area was used as counter electrode. All potentials values given in this work were referred on the SCE electrode. Electrochemical measurements were performed in aerated NaCl 3 wt % solution, at 298 ± 1 K.

The gravimetric measurements were performed under ambient conditions using an analytical balance (precision ±0.1 mg). The dimension of the rectangular copper coupons was 31 × 8 × 8 mm. After weighing, the treated coupons were suspended in our saline solution for different exposure periods (3 – 12 days). Then, the surface of the specimens was carefully washing and degreased with ethanol and acetone, dried at room temperature and then weighted. Triplicate experiments were carried out in each case and the mean weight losses were reported.

3. Results and discussion

3.1 Characterization of silane film formed on copper

In order to demonstrate the interaction mechanisms between silane coupling agents and substrate, the treated surface is submitted to: (i) only dipped specimens and dried at room temperature for 1440 min followed by washing with ethanol for 10 min (aged/Cu-silane) and (ii) heat treated at 100°C for 60 min followed by washing with ethanol for 10 min (cured/Cu-silane).

Figure 2 shows the FT-IR spectra of a pure γ-MPS (a) and the DRIFT of aged/Cu-MPS (b) and cured/Cu-MPS (c) in the region 4000-400 cm^{-1}. The band assignments for the pure γ-MPS are presented in **Table 1** [5].

It can clearly see that the spectrum of pure γ-MPS is different to the modified specimens (Cu-MPS) spectra.

In the spectra of aged/Cu-MPS and cured/Cu-MPS, the band around 2841 cm^{-1} corresponds to C-H stretching vibrations of the methoxy groups decrease and move to higher frequencies (2890 cm^{-1}). This designates a performs hydrolysis in the Si-OCH_3 linkages of the γ-methacryloxypropyltrimethoxysilane molecules attached on the substrate. These two spectra reveal an extensive band about 3427 cm^{-1} corresponded to OH stretching of Si-OH groups (hydrolysed Si-OCH_3) but only a decrease is detected in this band after cured treatment. The similar phenomenon is observed for Si-OH

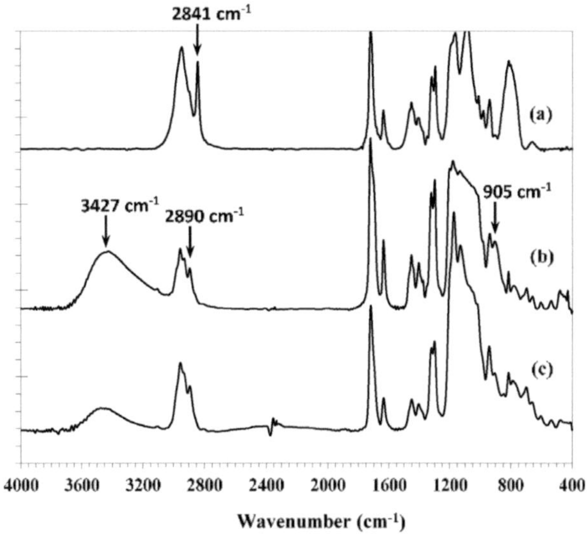

Figure 2.
FT-IR spectrum of pure γ-MPS (a) and DRIFT spectra of aged/Cu-MPS (b) and cured/Cu-MPS (c) (4000–400 cm^{-1} region).

Band position (cm^{-1})	Assignments	
2945	CH_3	ν_{as}(C-H)
2841	O-CH_3	ν_s(C-H)
1720[2]	methacrylic	ν(C=O)
1638	methacrylic	ν(C=C)
1454	CH_2	δ(C-H)
1404	Si-CH_2	δ(Si-C)
1320	C-O-C	ν_{as}(C-O)
1296	C-O-C	ν_s(C-O)
1191	O-CH_3	ρ(O-C)
1167	CH_3	δ(C-H)
1089	Si-O-C	ν_{as}(C-O)
1012	Si-O-Si	δ_{as}(Si-O)
980	Si-O-Si	δ_s(Si-O)
940	C=C-C=O	δ_s(C=C)
818	C=C-C=O	δ_{as}(C=C)

Table 1.
The FT-IR vibration band positions and their assignments of γ-MPS.

bond at 905 cm^{-1}. These evolutions prove the beginning of the silane condensation step. Moreover, the relatively higher Si-OH bands of uncured specimens, maybe due to the entrapped of water in the γ-MPS xerogel. A simulation of the spectrum region between 1770 and 1650 cm^{-1} (**Figure 3**) shows that after curing treatment, a decrease in C=O—HO bonding, giving a peak at around 1700 cm^{-1}, compared with the band at 1720 cm^{-1} assigned to $\nu_{C=O\text{-free}}$ [6]. This characterizes the interaction between –OH group (silanol and/or water) and C=O group.

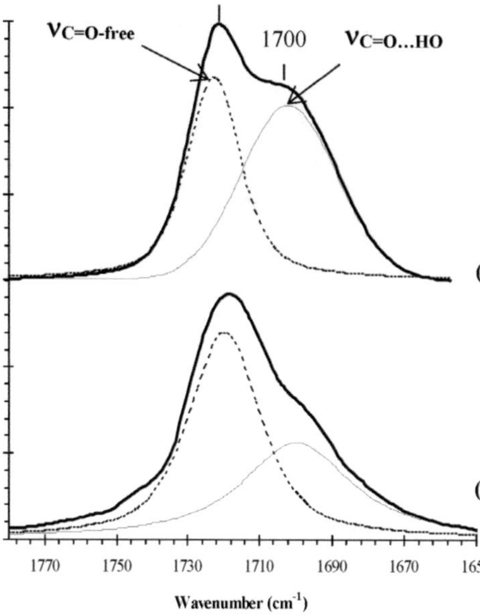

Figure 3.
C=O band deconvolution of Aged/Cu-MPS (a) and cured/Cu-MPS (b) (1770–1650 cm^{-1} region).

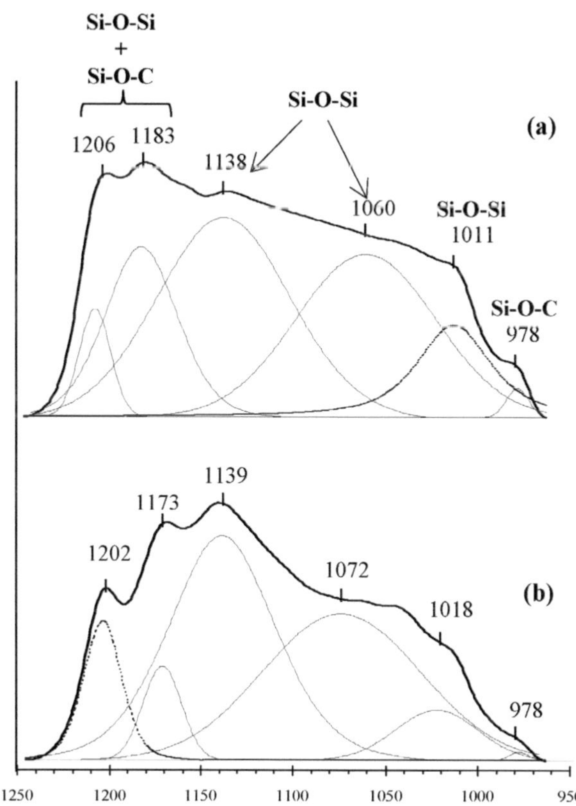

Figure 4.
Si-O band deconvolution of Aged/Cu-MPS (a) and cured/Cu-MPS (b) (1250–950 cm^{-1} region).

The bands characteristic of the different Si-O forms are observed in the region from 1250 cm^{-1} to 950 cm^{-1} in DRIFT spectral. A deconvolution of this region spectrum is used to elucidate the evolution of different Si-O forms (**Figure 4**). The presence of the peaks at around 1183 and 978 cm^{-1} indicates that the hydrolysis process remains unaffected (Si–OCH_3 remaining) [7]. The decreases of these bands, after heat curing (**Figure 4b**), recommend the auto-condensation of the methoxy groups and/or the grafting of coupling agent onto copper surface. As described in a previous research [8], the curing process can accelerate the condensation. The increase in the Si–O–Si bond, at around 1138 cm^{-1} and 1206 cm^{-1}, signify an intermolecular condensation between adjacent adsorbed –Si–OH groups. These results are in good agreements with those of other studies dealing with glass surfaces coated with the tow similar silanes [9]. Finally, **Figure 2b** and **c** also shows.

Finally, it can also be seen from **Figure 2b** and **c** the existence of several peaks in the region of the infrared spectrum from 850 cm^{-1} to 400 cm^{-1}, particularly after curing. This is may be caused by the Cu–O band [8, 10].

In the case of γ-APS modified copper, DRIFT spectra of the aged/Cu-APS (a) and cured/Cu-APS (b) are exhibited in **Figure 5**. For the reference, the FT-IR spectrum of modified substrate has been superposed in this figure. Significant changes in the intensity of the bands can be clearly seen in the range from 1250 to 850 cm^{-1} [11]. For the FT-IR spectrum of γ-APS (**Figure 5c**), the absorption bands at 1167, 1104, 1080 and 957 cm^{-1} related to Si-O-C_2H_5 group disappeared after silanization in an aqueous solution. The major IR absorption peaks and tentative assignments band for pure γ-APS are cited in our previously work [8].

A new broad band at around 1010–1140 cm^{-1}, characteristic of siloxane bonds, appeared instead in the DRIFT Spectra, signifying that most ethoxy groups were hydrolyzed and Si–O–Si linkages were formed on the surface by condensation and lateral polymerization of γ-APS. These absorption bands become clearer after curing treatment. This means that there is an evolution toward better interconnected

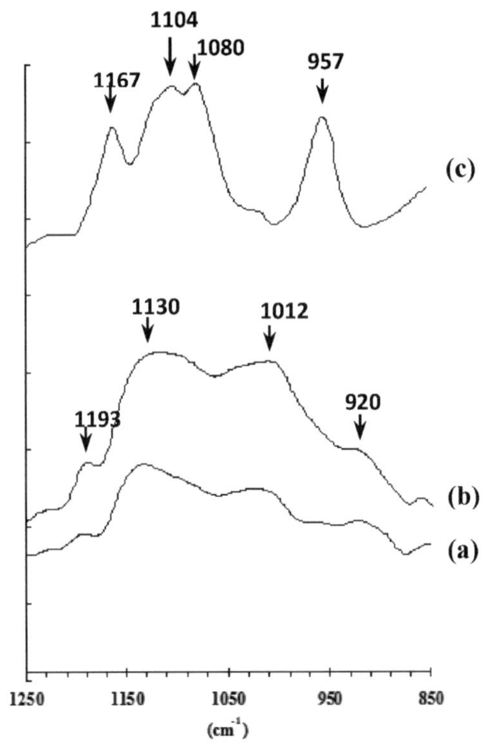

Figure 5.
Infrared spectra for fresh/Cu-APS (a), aged/Cu-APS (b), cured/Cu-APS (c) and aged-cured/Cu-APS (d) (850–400 cm^{-1} region).

networks [12, 13]. Contrary to the aging process, after curing this coupling agent adsorption via both physical and chemical interactions, so that chemisorption has been suggested to be the dominant adsorption mechanism. As previously reported [8, 11], the Si–O–Si stretching band at higher frequencies is due to cyclic siloxane units, and the lower one is assigned to the long Si–O–Si chains. After curing treatment, the proportion of the Si–O–Si chains peak (close to 1012 cm^{-1}) increases. This indicates that the higher temperature of treatment has a significant effect on the structure of γ-APS films. More Si–O–Si chains will be formed on the film and may contain fewer defects. Moreover, **Figure 5a** reveals the presence of a weak absorption band at around 920 cm^{-1} corresponding to the Si-O stretching of residual Si-OH groups. The intensity of this absorption band decreases sharply after heat treatment (**Figure 5b**), indicating additional polymerization of the adsorption γ-APS. Finally, the increase of Si-CH_2(CH_2CH_2) band at 1193 cm^{-1}, after heat treatment process, implies an increase in the silane quantity at the metal surface.

Finally, it can be concluded from infrared results, that the greater temperature of curing may favor both the surface condensation process (formation of Cu-O-Si bands in the interfacial layer through the reaction between Si–OH groups, and the hydroxylated copper surface) and the Si–O–Si linkages formation.

3.2 Corrosion protection

3.2.1 Potentiodynamic polarization

The polarization curves of pure copper (i.e., the blank) (a), aged/Cu-MPS (b), cured/Cu-MPS (c), aged/Cu-APS (b) and cured/Cu-APS (e) in aerated NaCl 3 wt % aqueous solution are shown in the **Figure 6**. It is well known that the cathodic part of the polarization curves can be divided into three regions of potential: Region I corresponds to the weak polarization region near OCP; Region II is associated to the reduction of dissolved oxygen (Eq. (6)); and Region III (below - 320 mV) is attributed to the hydrogen evolution rection given in Eq. (7) [14, 15]:

$$2H_2O + 2e^- \rightarrow 2OH^- + H_2 \quad (6)$$

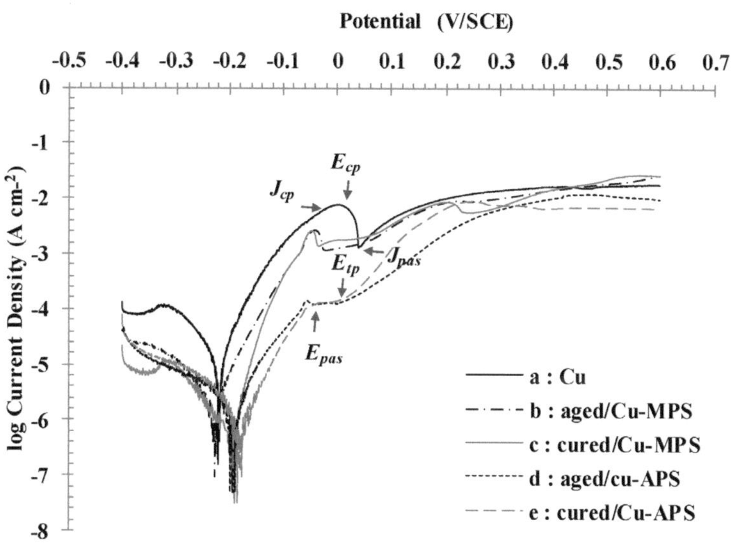

Figure 6.
Potentiodynamic polarization curves of Cu (a) aged/Cu-MPS (b), cured/Cu-MPS (c), aged/Cu-APS (d) and cured/Cu-APS (e), in NaCl 3 wt % solution.

$$O_2 + 2H_2O + 4e^- \rightarrow 4OH^- \tag{7}$$

It could be observed that the pH value of the solution after the electrochemical experiments is a little higher than that before the experiment.

The anodic part of the curve copper shows three main regions to [15, 16]:

In the first region, an apparent Tafel behavior is observed and the current density increases up to a critical passivation value (J_{cp}), due to the oxidation of Cu (0) to Cu(I) (Eq. (8))

$$Cu \rightarrow Cu^+ + e^- \tag{8}$$

In the second region, The current density decease from J_{cp} to full passivation current density (J_{pas}). Cu(I) is rapidly transformed to an insoluble CuCl film (Eq. (9)) [17].

$$Cu^+ + Cl^- \rightarrow CuCl \tag{9}$$

In aqueous solution, $CuCl$ would be unstable. It is immediately converted to the soluble cuprous complex $CuCl_2^-$ (Eq. (10)) [18]. The current density suddenly increases again from J_{pas} to elevated anodic potential. We detected a small full passivation field (7 mV) in the case of the untreated copper. Thus, the dissolution of substrate is happening step by step.

$$CuCl + Cl^- \rightarrow CuCl_{2\ (surface)}^- \tag{10}$$

In the third region, the potential increases again before to stabilize when it reaches a limiting current density. $CuCl_2^-$ is oxidized to Cu^{2+} ions according to the following reactions (Eqs. (11) and (12)):

$$CuCl_{2\ (surface)}^- \rightarrow CuCl_{2\ (solution)}^- \tag{11}$$

$$CuCl_{2\ (solution)}^- \rightarrow Cu^{2+} + 2Cl^- + 2e^- \tag{12}$$

However, according to some authors, other corrosion products could be also formed, such as Cu_2O, CuO, $Cu(OH)_2$ [14, 19–22].

Figure 6 also shows that the specimen coated with silane coupling agents have a different behavior for untreated copper. In addition, both cathodic and anodic current decreased and the corresponding slopes vary, which indicated that the γ-MPS or γ-APS coating is mixed-type corrosion coating, i.e. inhibitor and barrier coating.

The electrochemical parameters such as E_{corr}, corrosion potential, J_{corr}, corrosion current density, β_c, cathodic and β_a, anodic slopes, R_p, polarization resistance, CR, corrosion rate and P_{EF}, protective efficiency are calculated according to polarization curves and summarized in **Table 2**. The values of J_{corr} and E_{corr} were obtained by the extrapolation of anodic and cathodic Tafel curves. The R_p values are calculated by the next formula (Eq. (13)) [18],

$$R_p = \frac{B}{J_{corr}} \tag{13}$$

here, B is a constant that is calculated by using Stern–Geary equation (Eq. (14)) [23],

$$B = \frac{\beta_c \beta_a}{2.303(\beta_c + \beta_a)} \tag{14}$$

Solution	Parameters						
	E_{corr} (mV)	J_{corr} (μA cm^{-2})	-β_c (mV/dec)	β_a (mV/dec)	R_p (kΩ cm^2)	*CR* (mmy)	P_{EF} %
Cu	−220	17.78	104	58	1.07	0.207	—
aged/Cu-MPS	−224	2.85	103	55	5.46	0.033	84.06
cured/Cu-MPS	−195	0.63	82	30	15.14	0.007	96.62
aged/Cu-APS	−192	3.31	270	85	8.48	0.038	81.64
cured/Cu-APS	−180	2.14	150	70	9.68	0.025	87.86

Table 2.
Electrochemical kinetics parameters and protective efficiency obtained from potentiodynamic polarization curves.

The values of corrosion rate (*CR*, millimeters per year (mmy^{-1})) are calculated using the expression (Eq. (15)) [8],

$$CR = 3.268 \times 10^3 \frac{J_{corr}}{\rho} \frac{MW}{Z} \tag{15}$$

where ρ is the density of Cu in g.cm^{-3} (= 8.92), MW is molecular weight of copper in g and Z is the number of electrons transferred in the corrosion reaction; Z = 2 in the case of Cu reaction.

After silane-modified copper surface, the potentiodynamic polarization curves moved toward anodic direction and toward less current density. The values of corrosion potential shift in the positive direction, denoting the beneficial effect of the two coupling agents' treatments on copper substrate corrosion. These two treatments produce best results mainly on the copper anodic oxidation reaction when the coating is the γ-APS, whose currents are reduced by about two orders of magnitude, at 0 mV/SCE. Nevertheless, the cathodic currents are reduced only by about one order of magnitude for the tow used coupling agents.

Table 2 also depicts that the values of J_{corr} and *CR* decease, after treatments, while the protective efficiency increases sharply to reach 96.62% for the cured/Cu-MPS specimen). By leaving the silane coupling agent coating in contact with the substrate for 24 h, Si–O–Si linkages formation begins to take place. This generates a more robust and adherent coating. The notable hindrance to the copper anodic oxidation process, watched from the noticeable diminution in corrosion current density and corrosion rate values and the increase in the polarization resistance and the protective efficiency, can mostly be ascribed to the strong Si–O–Si linkages [11]. The corrosive attack can be demonstrated only in the coating holes and after a period of time it causes a noticeable attack of the metal. Additionally, heat treatment helps the interconnected networks of the coupling agent on the surface through the elimination of water molecules [24]. More condensation takes place mostly in the outermost part of the silane film, leading to a polymolecular, denser, less permeable and, consequently, more corrosion resistant coating.

Figure 6 also reveals that the untreated copper immersed in saline solution reaches passivity in a typical active-passive transition. Coupling agent modified copper in NaCl 3 wt % aqueouse solution features much wider potential range of full passivation, which offers further protection at elevated values of positive potential. The electrochemical parameters such as J_{cp}, critical current density of passivation, E_{cp}, critical passivation potential, J_{pas}, full passivation current density, E_{pas}, full passivation potential, E_{tp}, trans-passivation potential and ΔE, passivation range are listed in **Table 3**. Compared with the untreated copper (i.e., the blank), the E_{pas} and E_{cp} values shifted toward the cathodic direction and the J_{pas} and J_{cp} values decrease abruptly. The move of E_{pas} and E_{cp} values toward cathodic potentials demonstrates higher corrosion resistance of the coupling agent modified copper. Moreover, the lower value of full passivation current density verifies a better tightness on the silane film formed on the copper surface after treatments.

The passivation ability of the substrate corresponding to the significant criterion of passivation kinetics is verified through the drop of J_{cp} value. Indeed, the area delimited by the peak of activity, identified by J_{cp}, corresponds to the amount of electricity required for the passivation of the material. The weak dissolution of the material is, on its side, verified by the decrease of this surface which is easily passive. These phenomena become more pronounced when the used silane is γ-APS.

In the fully passive range, the current density is independent of the potential. When the potential of full passivation finishes, E_{tp} is reached. After this, the passive film continuity is damaged and the metal gets trans-passivated.

The passivation of silane-modified copper becomes more rapid and the passivation range, characterized by the difference between E_{tp} and E_{pas} ($\Delta E = E_{tp} - E_{pas}$), is more extensive than that of pure copper. This confirms the relatively high corrosion resistance of silane modified copper.

At the end of the passivation, an unexpected increase in the current density has been detected, along with the evolution of molecular oxygen as well as the return of the active area, where copper dissolution takes place.

3.2.2 Gravimetric measurements

To investigate the effect of the treatments on the corrosion inhibition of copper in aerated NaCl 3 wt % aqueous solution at ambient temperature, gravimetric measurements were carried out. **Figure 7** shows the plot of the weight losses versus time curves of specimen blank (a) aged/Cu-MPS (b), cured/Cu-MPS (c), aged/Cu-APS (d) and cured/Cu-APS (e). The weight loss (Δm, mg cm^{-2}) and the corrosion rate (CR, mg cm^{-1} h^{-1}) were calculated as follows (Eqs. (16) and (17)) [25]:

$$\Delta m = \frac{W_1 - W_2}{A} \tag{16}$$

$$CR = \frac{\Delta m}{t} \tag{17}$$

Where, W_1 and W_2 are the weight before and after exposure to saline solution, respectively, A is the total surface area and t is the immersion period.

According to **Figure** 7, the weight losses of untreated copper in NaCl 3 wt % solution increases with the increase in the immersion period as a result of the

Solution	**Parameters**					
	E_{cp} (mV)	**J_{cp} (μA cm^{-2})**	**E_{pas} (mV)**	**J_{pas} (μA/cm^2)**	**E_{tp} (mV)**	**ΔE (mV)**
Cu	7	7586	40	1288	47	7
aged/Cu-MPS	−41	2690	−25	1170	40	15
cured/Cu-MPS	−43	2570	−30	1410	46	16
aged/Cu-APS	−57	138	−46	126	−3	43
cured/Cu-APS	−55	105	−47	112	−2	45

Table 3.
Characteristic passivation parameters obtained from potentiodynamic polarization curves.

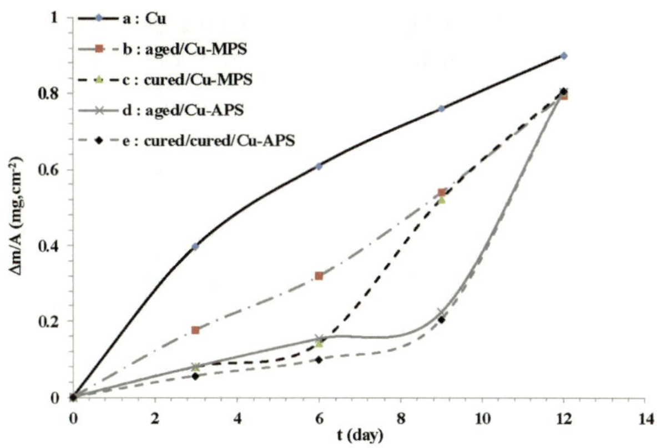

Figure 7.
Variations of the weight losses with time for Cu (a) aged/Cu-MPS (b), cured/Cu-MPS (c), aged/Cu-APS (d) and cured/Cu-APS (e), in NaCl 3 wt % solution.

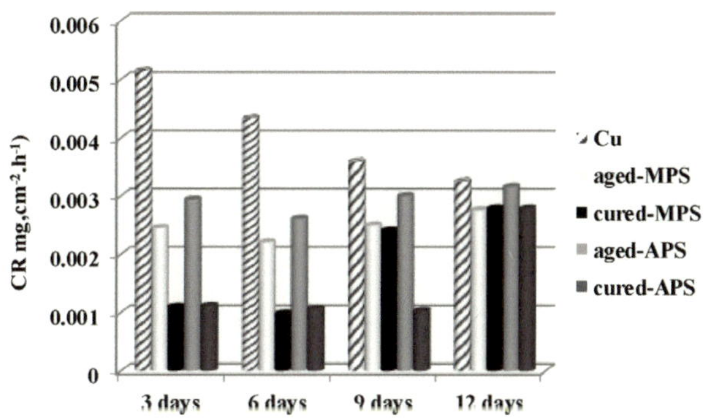

Figure 8.
Variations of the corrosion rate with time the different treatments.

continuous dissolution of copper ions, due to the severe aggressiveness of the chloride ions.

aged/Cu-MPS decreases the weight loss of specimen, especially after 3 and 6 immersion days. Above this period, the weight loss value increases significantly to reach the value obtained for blank specimen after 12 immersion days. A further decrease in the loss of weight took place after curing and/or when the coating is γ-APS. The advantageous effect endures until 9 days of immersion, which indicates an improvement in coupling agent film durability. In fact, curing treatment catalyzes the formation of both chemisorbed and polymerized product forming more robust coupling agent layer. Nevertheless, after 12 immersion days this layer cannot resist whatever the silane coating used. This is also illustrated by the **Figure 8**, from which it can be seen that, after this period, there is no notable difference of the corrosion rate obtained for the untreated or silane coated copper.

4. Conclusions

The main conclusions of this study are summarized below:

- Room temperature aging allows a certain condensation between the loosely adsorbed silane molecules, which relatively improves the performances of the silanic layer.

- Curing treatment may favor both the surface condensation process (formation of Cu-O-Si bands in the interfacial layer through the reaction between silanol groups and the hydroxylated copper surface) and the siloxane band formation.

- γ-MPS and γ-APS act as mixed corrosion inhibitor for copper in NaCl 3 wt % solution.

- The corrosion protection offered by coupling agent coating is due to both the blocking of copper surface parts with the reduction of oxygen and the metal dissolution occurring in the pores of the coating layers.

- A relatively high corrosion resistance of γ-MPS modified copper, especially after curing treatment, was remarked. Therefore, the passivation zone is more extensive when the coating is γ-APS.

- Gravimetric results show that the curing process leads to lower weight loss for the silane film, chemisorbed at the copper surface, after 3 and 6 immersion days. However, after 12 days of immersion this film cannot resist whatever the silane used.

Author details

Mohamed Masmoudi
Laboratory of Electrochemistry and Environment (LEE), Sfax National Engineering School, Sfax Preparatory Engineering Institute, University of Sfax, Sfax, Tunisia

*Address all correspondence to: med_mmasmoudi@yahoo.fr

References

[1] Van Ooij WJ, Zhu D, Stacy M, Seth A, Mugada T, Gandhi J, Puomi P. Corrosion Protection Properties of Organofunctional Silanes-An Overview. Tsinghua Science & Technology. 2005; 10:639-664.

[2] Subramanian V, Van Ooij WJ. Silane based metal pretreatments as alternatives to chromating:Shortlisted. Surface Engineering. 1999;15:168-172.

[3] Cecchetto L, Denoyelle A, Delabouglise D, Petit JP. A silane pre-treatment for improving corrosion resistance performances of emeraldine base-coated aluminium samples in neutral environment. Applied Surface Science. 2008;254:1736-1743.

[4] Arkles B, Steinmetz JR, Zazyczny J, Mehta P. Factors contributing to the stability of alkoxysilanes in aqueous solution.Journal of Adhesion Science and Technology. 1992;6:193-206.

[5] Masmoudi M, Abdelmouleh M, Abdelhedi R. Infrared characterization and electrochemical study of γ-methacryloxypropyltrimethoxysilane grafted in to surface of Copper. Spectrochimica Acta Part A: Molecular and Biomolecular Spectroscopy. 2014; 118:643-650.

[6] Abdelmouleh M, Boufi S, Ben Salah A, Belgacem, MN Gandini. Interaction of Silane Coupling Agents with Cellulose. Langmuir. 2002;18: 3203-3208.

[7] Kozelj M, Vuk AS, Jerman I, Orel B. Corrosion protection of Sunselect, a spectrally selective solar absorber coating, by (3-mercaptopropyl) trimethoxysilane. Solar Energy Materials & Solar Cells. 2009;93:1733-1742.

[8] Masmoudi M, Rahal C, Abdelmouleh M, Abdelhedi R. Hydrolysis process of γ-APS and characterisation of silane film formed on copper in different conditions. Applied Surface Science. 2013;286:71-77.

[9] Li C, Yin Y, Hou H, Fan N, Yuan F, Shi Y, Meng Q. Preparation and characterisation of $Cu(OH)_2$ and CuO nanowires by the coupling route of microemulsion with homogenous precipitation. Solid State Communications. 2010;150: 585-589.

[10] Yan X, Xu G. Surface Coating Technology. Synergy effect of silane and CTAB on corrosion-resistant property of low infrared emissivity Cu/ polyurethane coating formed on tinplate. 2010;204:1514-1520.

[11] Rahal C, Masmoudi M, Abdelmouleh M, Abdelhedi R. An environmentally friendly film formed on copper: characterization and corrosion protection. Progress in Organic Coatings. 2015;78:90–95.

[12] Ishida H, Naviroj S, Tripathy SK, Fitzgerald JJ, Koenig JL, The Structure of an Aminosilane Coupling Agent in Aqueous Solutions and Partially Cured Solids. Journal of Polymer Science. 1982; 20:701-718.

[13] Song J, Van Ooij WJ. Bonding and corrosion protection mechanisms of γ-APS and BTSE silane films on aluminum substrates. Journal of Adhesion Science and Technology. 2003;16:2191-2221.

[14] Liao X, Cao F, Zheng L, Liu W, Chen A, Zhang J, Cao C. Corrosion behaviour of copper under chloride-containing thin electrolyte layer. Corrosion Science. 2011;53:3289-3298.

[15] Masmoudi M, Rahal C, Abdelhedi R, Khitouni M, Bouaziz M. Inhibitive action of stored olive mill wastewater (OMW) on the corrosion of copper in a NaCl solution. RSC Advances. 2015;5: 101768-101775.

[16] Otmacic H, Lisac ES. Copper corrosion inhibitors in near neutral media. Electrochimica Acta. 2003;48: 985-991.

[17] Refait P, Rahal C, Masmoudi M. Corrosion inhibition of copper in 0.5 M NaCl solutions by aqueous and hydrolysis acid extracts of olive leaf. Journal of Electroanalytical Chemistry. 2020;859:113834-113843.

[18] Rahal C, Masmoudi M, Abdelhedi R, Sabot R, Jeannin M, Bouaziz M, Refait P. Olive leaf extract as natural corrosion inhibitor for pure copper in 0.5 M NaCl solution: A study by voltammetry around OCP. Journal of Electroanalytical Chemistry. 2016;769: 53–61.

[19] Sherif EM, Erasmus RM, Comins JD. Corrosion of copper in aerated synthetic sea water solutions and its inhibition by 3-amino-1,2,4-triazole. J. Colloid Interface Science. 2007;309:470-477.

[20] Chu C, Lee C, Wang Y, Wan C, Chen C. The role of cuprous ion as corrosion inhibitor for copper in a chloride medium. Journal of the Chinese Institute of Chemical Engineers. 2007; 38:361-364.

[21] Scendo M. Inhibition of copper corrosion in sodium nitrate solutions with nontoxic inhibitors. Corrosion Science. 2008;50:1584-1592.

[22] Kear G, Barker BD, Walsh FC, Electrochemical corrosion of unalloyed copper in chloride media–a critical review. Corrosion Science. 2004;46: 109-135.

[23] Zhong X, Li Q, Hu J, Yang X, Luo F, Dai Y. Effect of cerium concentration on microstructure, morphology and corrosion resistance of cerium–silica hybrid coatings on magnesium alloy AZ91D. Progress in Organic Coatings. 2010;69:52-56.

[24] Doidjo MRT, Belec L, Aragon E, Joliff Y, Lanarde L, Meyer M, Bonnaudet M, Perrin FX. Silane Coupling Agent for Attaching Fusion-Bonded Epoxy to Steel. Progress in Organic Coatings. 2013;76:1765-1777.

[25] Sherif EM, Park SM. 2Amino5-ethyl-1,3,4-thiadiazole as a corrosion inhibitor for copper *in* 3.0% NaCl solutions. Corrosion Science. 2006;48; 4065-4079.

Chapter 3

Fabrication and Characterization of Cobalt-Pigmented Anodized Zinc for Photocatalytic Application

Judith Chebwogen and Christopher Mkirema Maghanga

Abstract

Population growth and urbanization have led to water scarcity and pollution, which is a health hazard not only to humans but also to the ecosystem in general. This has necessitated coming up with ways of treating water before consumption. Photocatalysis has proved to be one of the most promising cheap techniques that involve chemical utilization of solar energy. TiO_2 widely used in photocatalysis absorbs a narrow range of the solar spectrum compared to *ZnO*. In this regard, this study aimed at preparing and optimizing cobalt-pigmented *ZnO,* which is applicable in photocatalytic water treatment. The objectives in this study were to fabricate zinc oxide (*ZnO*) thin films by anodization, pigment the fabricated films with varying cobalt concentrations, characterize the fabricated films optically, and investigate the cobalt-pigmented *ZnO*'s performance in the methylene blue degradation under UV light irradiation. Mirror-polished zinc plates were sonicated in ethanol and rinsed. Anodization was done at room temperature in 0.5 M oxalic acid at a constant voltage of 10 V for 60 min, and cobalt electrodeposited in the films. Post-deposition treatment was done at 250°C. Optical properties of the films were studied using a UV-VIS-NIR spectrophotometer in the solar range of 300–2500 nm. The photocatalytic activity of the fabricated films was studied in methylene blue solution degradation in the presence of UV light irradiation for 5 h. Cobalt pigmenting was observed to reduce reflectance and optical band gap from 3.34 to 3.10 eV indicating good photocatalytic properties. In this study, *ZnO* film pigmented with cobalt for 20 s was found to be the most photocatalytic with a rate constant of 0.0317 h^{-1} and hence had the optimum cobalt concentration for photocatalytic water treatment. This can be applied in small-scale water purification.

Keywords: photocatalysis, anodization, zinc oxide, pigmenting, optical properties

1. Introduction

Contaminated waste water from industries, refineries, agriculture run-off, domestic and sewage water contain organic pollutants that are not only harmful to human but also the ecosystem in general. This calls for the need to treat water before discharge to the environment. Some of the water treatment methods used are reverse osmosis, sedimentation, filtration, distillation, coagulation and flocculation, chlorination, photocatalysis and aeration. According to Ref. [1], photocatalytic

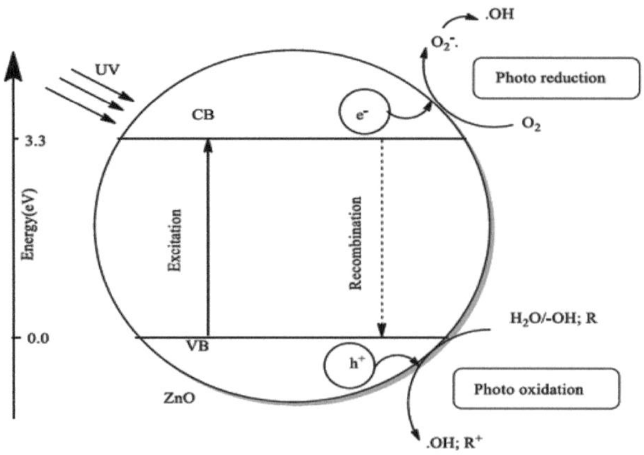

Figure 1.
Photocatalyst e− and h+ generation and their possible reactions in aqueous solutions [2].

water treatment has proved to be one of the most promising ways of purifying waste water. This technique degrades the hazardous organic pollutants in water by chemically utilizing solar energy. It involves the excitation of electrons from the valence band to the conduction band of a semiconductor by a radiation with sufficient energy. The photogenerated electrons and holes react with oxygen and hydrogen in water respectively to form superoxide and hydroxyl radicals as illustrated in **Figure 1**.

Photocatalytic water treatment is preferred because it is efficient, simple, possible to use solar light thus reusable and also capable to mineralize pollutants into products which are environment friendly [3]. Photocatalysis is also applied in air purification, in antibacterial action, hydrogen evolution through water splitting, designing self-cleaning surfaces, sterilization and photo electrochemical conversion. In these applications, semiconductors are used as photocatalysts.

Some of the attractive photocatalysts applied in photocatalytic degradation are Titanium dioxide (TiO_2), zinc oxide (ZnO), tin oxide (SnO_2). Iron (iii) oxide (Fe_2O_3), Vanadium (v) oxide and niobium pentoxide (Nb_2O_5) have also been reported by [4]. TiO_2 is the most studied and applied in photocatalysis because of its structural stability, nontoxicity and effectiveness according to [5, 6]. Its only drawback as compared with ZnO is that it absorbs a narrow range of the solar spectrum yet they have similar band gap energies, [7]. In addition, ZnO is readily available and cheaper than TiO_2.

2. Zinc oxide (ZnO) properties and applications

ZnO is an n-type direct wide band gap semiconductor whose crystals exist in hexagonal wurtzite, zincblende and rock salt structures but the most common structure is hexagonal wurtzite because it is stable. Its band gap energy is about 3.37 eV at room temperature, and it has a high exciton energy of 60 meV.

According to [8], *ZnO* photocatalysis is influenced by its direct band gap and large free exciton binding energy (60 meV). ZnO also possess unique properties like photosensitivity, nontoxicity, environmental stability and good optical and electrical properties. These properties according to [9] have attracted much interest in varied applications. It also has high electron mobility, good thermal conductivity and transparency. All these properties make *ZnO* a promising material in several semiconductor applications including designing of transparent thin film transistors, light emitting diodes (LEDs), heat protecting windows and transparent electrodes in liquid crystal displays. They are also vital in applications like antibacterial use as reported by [10], dye sensitized solar cells by [11] gas sensors by [12] and photocatalysis as reported by [13].

2.1 ZnO photocatalytic water treatment

In photocatalytic water treatment, light energy greater than that of ZnO band gap illuminated on its surface excites electrons from its valence band to the conduction band leaving behind holes in the valence band. The excited electrons then interact with the surrounding oxygen to form superoxide radicals $\left(O_2^{-*}\right)$ while holes interact with water to form hydroxyl radicals (HO^*). These radicals react with the organic compounds dissolved in water decomposing them to water and carbon (IV) oxide. Photocatalytic water treatment process can be summarized using the equations, [14]:

$$ZnO + hv \rightarrow e_{cb}^- + h_{vb}^+ \tag{1}$$

$$h_{vb}^+ + H_2O \rightarrow H^+ + HO^* \tag{2}$$

$$e_{cb}^- + O_2 \rightarrow O_2^{-*} \tag{3}$$

$$O_2^{-*} + H^+ \rightarrow HO_2^* \tag{4}$$

$$HO_2^* + HO_2^* \rightarrow H_2O_2 + O_2 \tag{5}$$

$$H_2O_2 + e_{cb}^- \rightarrow HO^* + HO^- \tag{6}$$

$$H_2O_2 + hv \rightarrow 2HO^* \tag{7}$$

$$R + HO^* \rightarrow CO_2 + H_2O \tag{8}$$

Figure 2 illustrates how ZnO photocatalysis degrades organic pollutants.

ZnO photocatalytic activity is limited by its wide band gap property which makes it active only in the UV region of the solar spectrum [15]. This can be enhanced by doping or pigmenting it with a transition metal or nonmetal where the impurity added reduce the band gap consequently expanding its response to solar radiation. The preferred metal for pigmenting ZnO is Cobalt which has an ionic radius (0.745 Å) close to that of Zinc (0.74 Å) as stated by [16].

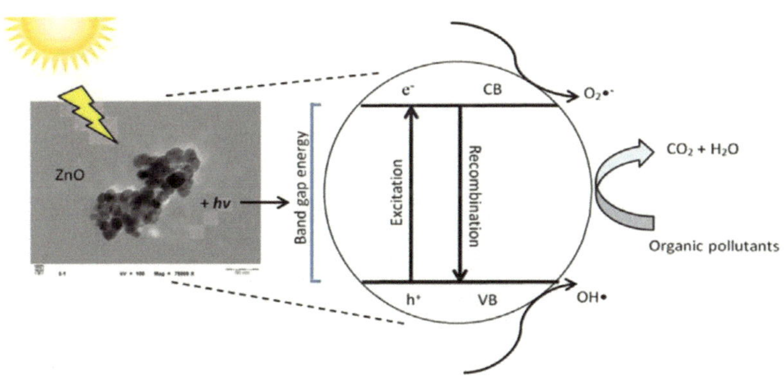

Figure 2.
Diagram showing ZnO *photocatalytic mechanism [13].*

3. Fabrication of zinc oxide thin films and nanostructures

Several techniques have been adopted in ZnO synthesis into different morphologies. These include spray pyrolysis, hydrothermal method, Radio Frequency (RF) and Direct Current (DC) magnetron sputtering, sol-gel and electrochemical anodization. The technique used determines the ZnO nanostructure formed. These nanostructures can be nanowires, nanoflowers, nanobelts and nanoparticles. ZnO thin films can also be fabricated using these techniques.

The choice of the technique in designing ZnO thin films is determined by the quality of the films required, simplicity, fabrication time and cost among others. In the synthesis of ZnO thin films, [17] acknowledged electrochemical anodization as a simpler route to design thin films. This technique is cheap. Simple and affordable.

3.1 Anodization

This is a simple two electrode configuration which involves a working electrode (anode) and a counter electrode (cathode) dipped in an electrolyte. A constant direct current passed through the electrodes results in the formation of a thin oxide layer on the surface of certain metals. The schematic diagram is as illustrated in **Figure 3**.

In this method of fabrication, several parameters like electrolyte temperature, inter-electrode spacing, electrolyte concentration, applied voltage and anodizing time should be controlled since they not only affect the nanostructure formation but also the surface density. While the type of electrolyte used determine the shape of the nanostructure, electrolyte concentration, anodizing time and the applied voltage influence the nanostructure density.

In the formation of ZnO thin films by anodization, Zinc metal is the working electrode (anode) and any inert material is the counter electrode (cathode). A constant current is passed from a power supply through the electrodes which results in redox reactions. These reactions are expressed as:

At the cathode:

$$2H^{+} + 2e^{-} \rightarrow H_2 \tag{9}$$

At the anode:

$$\mathrm{Zn} \rightarrow Zn^{2+} + 2e^{-} \tag{10}$$

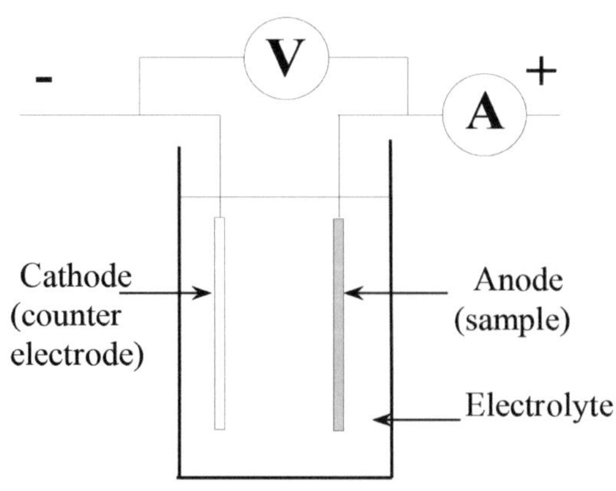

Figure 3.
Schematic diagram for the experimental setup for anodization [18].

$$Zn^{2+} + 2OH^- \rightarrow Zn(OH)_2 \tag{11}$$

$$Zn(OH)_2 \rightarrow ZnO + H_2O \tag{12}$$

4. Experiment

ZnO thin films were fabricated by anodizing about 99% pure ultrasonically cleaned Zinc metal in 0.5 M magnetic stirred Oxalic acid electrolyte at room temperature (300 K) for 60 min. Electric current was passed through the zinc metal (working electrode) and graphite (counter electrode) at a constant voltage of 10 V. The anodized films were rinsed in distilled water and then left to dry in air. This led to the formation of a white anodic film on the surface of Zinc metal. The experiment setup for anodizing Zinc is shown in **Figure 4**.

Varying Cobalt concentrations were incorporated into ZnO films by electrodeposition method using 0.5 M Cobaltous (II) sulphate solution as the Cobalt source. This was done by passing alternating current from a 20 V source through the Cobaltous (II) sulphate solution containing ZnO working electrode and graphite counter electrode for 10, 20, 30, 40, 50 and 60s. The films were then rinsed and left to dry in air. Heat treatment was done in a Carbolite 301 temperature controller at 523 K for 2 h.

The amount of Cobalt deposited in ZnO films at different deposition times was obtained using the Faraday's law of electrochemistry which is expressed mathematically as [19]:

$$m = \left(\frac{Q}{F}\right)\left(\frac{M}{z}\right) \tag{13}$$

where m is the mass of Cobalt deposited expressed in grams, Q is the charge in coulombs passed through ZnO, F is the Faraday constant (96485.33289 $Cmol^{-1}$), M represents the molar mass of Cobalt expressed in grams mol^{-1} and z is the valency number of Cobalt. Q is given by the product of current passed and time in seconds ($Q = It$) where the current passed is 2.18 A.

Optical characterization was done using PERKIN ELMER UV/VIS/NIR Lambda 19 spectrophotometer equipped with an integrating sphere which can be used to measure the absorbance, transmittance and reflectance of the films.

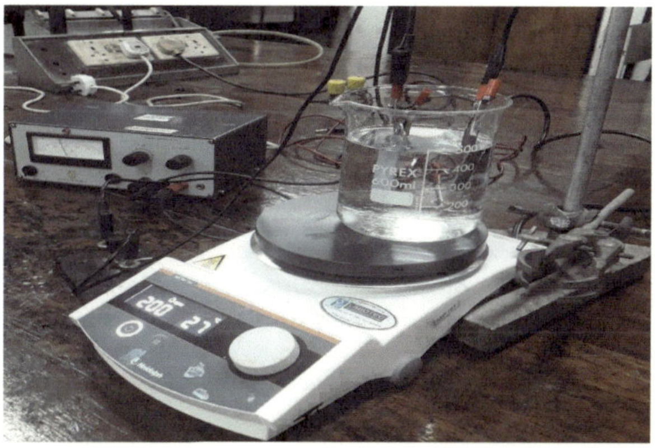

Figure 4.
Plate showing the experimental setup for anodizing zinc metal.

This spectrophotometer was set to measure the reflectance of the films in the solar range 300 nm < λ < 2500 nm against a barium sulphate reference standard.

The photocatalytic activity of unpigmented and Cobalt-pigmented *ZnO* thin films was studied in the degradation of aqueous methylene blue solution which is the simulated pollutant under UV light. The UV source composed of a UV cabinet with a UV lamp of wavelength 366 nm, 2 × 6 W irradiation power and methylene blue absorbance was measured using Optima SP-3000nano UV-VIS spectrophotometer. 60 ml of methylene blue with 10 ppm concentration was put in a petri dish and placed in the dark for 60 min to allow for adsorption-desorption equilibrium.

The petri dish and its contents were then transferred into the UV cabinet and illuminated for 5 h at ambient temperature. 1 ml of the degrading solution was drawn at 30 min intervals and its absorbance directly measured at 664 nm using the UV-VIS spectrophotometer. The absorbance recorded in this case was directly proportional to methylene blue concentration. This was revealed by a calibration curve plotted from standard methylene blue solutions of known concentrations prepared prior to the measurement. The calibration curve was used to obtain the concentration of methylene blue in the experiment with respect to absorbance.

4.1 Data analysis

Data was analyzed using SCOUT software which allows for analysis and simulation of optical spectra such as reflectance, transmittance and absorbance. This was done by fitting the measured experimental data into the simulated data in the software with the aid of different models. In this case, the models used are the Drude model for free carriers, harmonic oscillator to describe the atomic microscopic vibrations and Tauc Lorentz model to determine the band gaps of the films.

5. Results

A white ZnO thin film formed on Zinc electrode surface as a result of the redox reactions at the anode and cathode. The average film thickness as obtained from fitting of the experimental measured spectra into the simulated spectra using SCOUT software was 110 nm. **Figure 5** shows one of the obtained fitted spectra.

5.1 Reflectance spectra

The reflectance spectra of the fabricated films obtained from the spectrophotometer is shown in **Figure 6**.

As seen in the **Figure 6**, ZnO reflectance was affected by Cobalt pigmenting because Cobalt pigmented ZnO films had a lower reflectance than unpigmented ZnO. This decrease in reflectance may be attributed to darkening of the films when Cobalt concentration was increased. Lowered reflectance may also be as a result of the films becoming rough as Cobalt is deposited according to Ref. [20]. The more the Cobalt concentration, the rougher the films hence the decrease in the quantity of reflected light implying increased light absorption (absorbance).

5.2 Absorption coefficient

The absorption coefficient of the fabricated films was obtained from reflectance data of the films using the relation [21]:

$$R + T = e^{-\alpha d} \tag{14}$$

which yields

$$\alpha = \frac{1}{d} \ln \left[\frac{1}{R(\lambda)} \right] \tag{15}$$

where α is the absorption coefficient, d is the film thickness and $R(\lambda)$ is reflectance as a function of wavelength. **Figure 7** shows the variation of the absorption coefficient of the films with wavelength.

From the figure, it was observed that there is a sudden increase in the absorption coefficient at shorter wavelengths about 348 nm. This peak corresponds to ZnO absorption edge indicating that ZnO absorbs at short wavelengths in the UV region of the solar spectrum. It was also observed that Cobalt pigmenting affected the absorption coefficient since an increase in the Cobalt concentration led to an increase in the absorption coefficient. This may be attributed to the decrease in

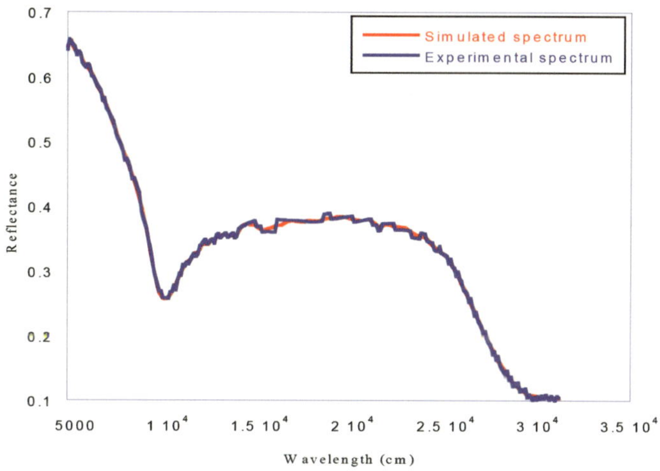

Figure 5.
Illustration of fitting of experimental to simulated spectra.

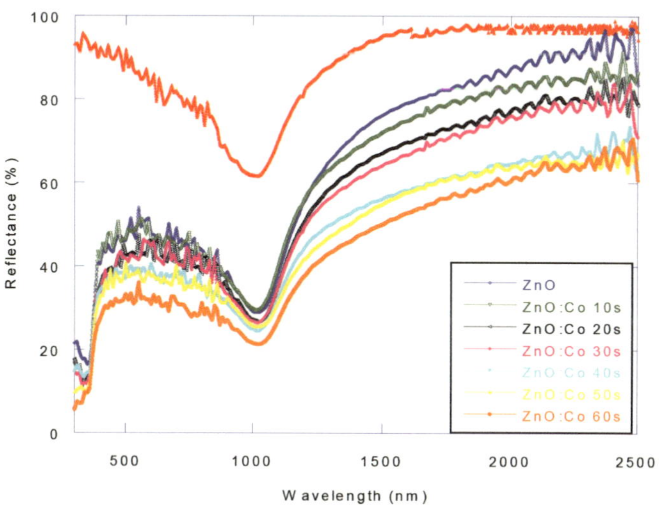

Figure 6.
Measured reflectance for zinc and the ZnO films with different cobalt concentrations.

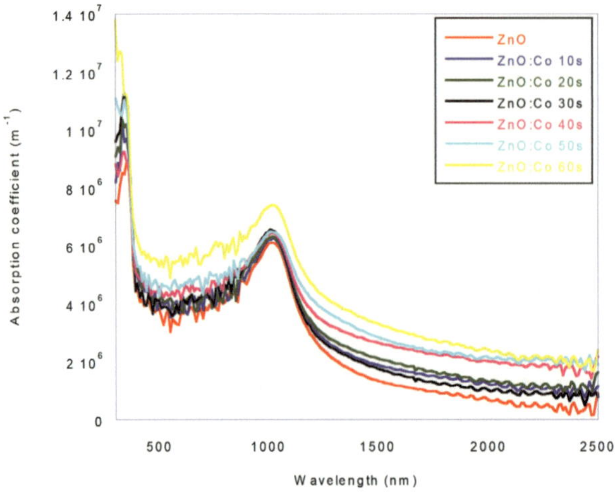

Figure 7.
Variation of absorption coefficient of the films with wavelength.

reflectance of the films as Cobalt concentration was increased. According to Ref. [22], the content of pigment in a film affects its absorption. Another peak was observed at about 1000 nm which shows absorption resulting from interband transition in the Zinc substrate [23].

5.3 Optical band gap

ZnO is a direct band gap semiconductor whose band gap can be obtained from a plot of $(\alpha h\nu)^2$ versus energy in eV. **Figure 8** shows the band gaps of the fabricated films.

The band gaps of the films ranged from 3.34 to 3.10 eV for pure ZnO and ZnO: Co 60s films respectively. The fabricated pure *ZnO* films had a band gap of 3.34 eV which is similar to the preceding work done in Ref. [24] who also found 3.34 eV. The gradual decrease observed as Cobalt was introduced into the *ZnO* films in different amounts is associated with the red shift of the absorption edge as observed in the reflectance spectra.

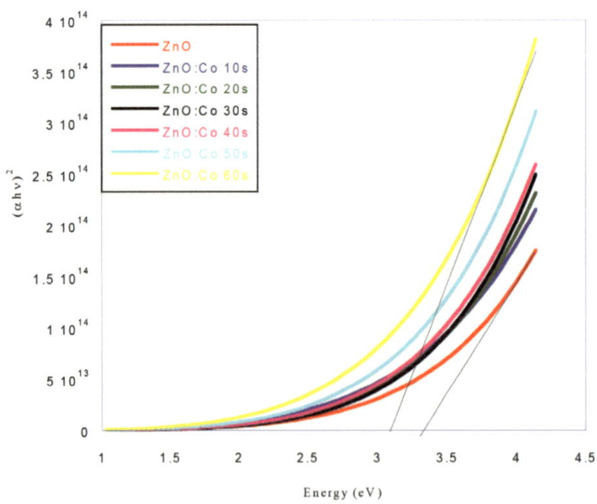

Figure 8.
Graph showing the band gap analysis of the fabricated films.

In Refs. [25–27], a gradual decrease in ZnO band gap with Cobalt pigmenting is also reported. They attributed the decrease to sp-d exchange interactions between the d electrons of Cobalt and *ZnO* conduction band electrons once the Cobalt ions substitute the Zinc ions in the crystal lattice. It was also stated that High Cobalt concentration led to the wavefunctions of the electrons in the Cobalt atoms overlapping as the Cobalt density increases resulting in the formation of an energy band by the overlapping forces consequently reducing the gap.

5.4 Photocatalytic degradation of methylene blue

In the investigation of photocatalytic degradation of methylene blue solution, the pseudo-first order kinetic model was used which according to [28] enables quantification of the photocatalytic activity of samples. This model involves the presentation of raw data as integral data. The performance of the pure and some of the Cobalt pigmented *ZnO* films was as shown in **Figure 9**. Methylene blue degradation without catalyst was used the control experiment.

From the figure, it is observed that the rate at which methylene blue with catalyst degraded was faster than that without the catalyst. Also, Cobalt pigmenting increased the degradation rate which may be attributed to the decrease in the optical band gap resulting from the red shift which was observed in the reflectance spectra. This shift resulted in more electrons gaining kinetic energy consequently moving to the conduction band to take part in the degradation process.

ZnO:Co 20s is had a higher rate of degradation with its degradation rate of 0.0317 h^{-1} obtained using the relation [28]:

$$-\ln\left(\frac{C}{C_0}\right) = kt \tag{16}$$

where k is the degradation rate constant given by the slope of the graph.

Heavy Cobalt content however lowered the methylene blue degradation rate. This may be attributed to the fact that excess metal pigment covered the active sites

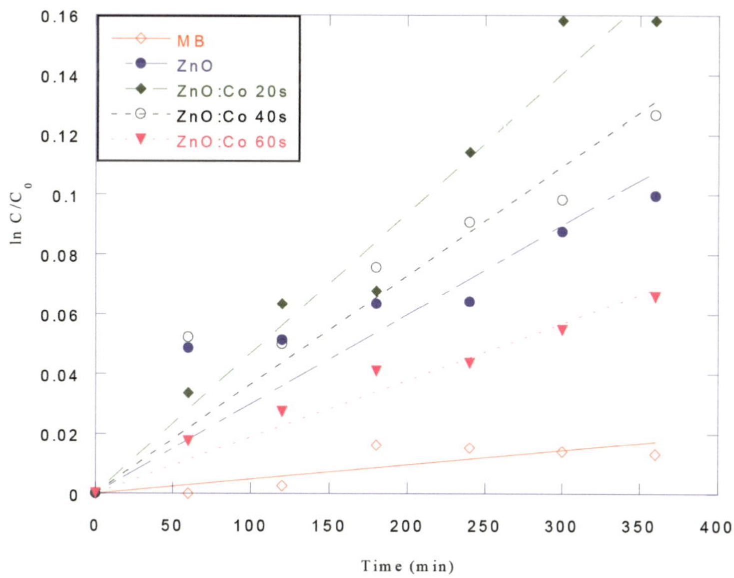

Figure 9.
A graph of ln (C/C_0) versus time in minutes for sampled films.

of the *ZnO* catalyst lowering its activity. It may also be due to the generation of the impurity levels deep in *ZnO* band gap which acted as recombination centers for the photogenerated electrons.

6. Conclusion

- ZnO thin films were successfully fabricated by anodization. The film formed was white in color.

- Reflectance of the ZnO films decreased with increase in cobalt content; hence, the increase in absorption on pigmenting is attributed to the films becoming rougher with increase in cobalt concentration.

- The band gap of ZnO changed from 3.34 eV for unpigmented to 3.10 eV for the most pigmented film.

Acknowledgements

The authors acknowledge the International Science Programme of Uppsala University who funded the research and the Departments of Chemistry and Physics of The University Of Zambia for their laboratory facilities under the auspices of Materials Science and Solar Energy Network for Eastern and Southern Africa (MSSEESA).

Author details

Judith Chebwogen* and Christopher Mkirema Maghanga
Kabarak University, Nakuru, Kenya

*Address all correspondence to: judithkoskey86@gmail.com

References

[1] Hoffmann MR, Martin ST, Choi W, Bahnemann DW. Environmental applications of semiconductor photocatalysis. Chemical Reviews. 1995; **1**:69-96

[2] Irani M, Mohammadi T, Mohebbi S. Photocatalytic degradation of methylene blue with ZnO nanoparticles; a joint experimental and theoretical study. Journal of the Mexican Chemical Society. 2016;**60**(4):218-225

[3] Prihod'ko RV, Soboleva NM. Photocatalysis: Oxidative processes in water treatment. Journal of Chemistry. 2013

[4] Vinu R, Madras G. Environmental remediation by photocatalysis. Journal of the Indian Institute of Science. 2012; **90**(2):189-230

[5] Fujishima A, Rao TN, Tryk DA. Titanium dioxide photocatalysis. Journal of Photochemistry and Photobiology C: Photochemistry Reviews. 2000;**1**(1): 1-21

[6] Klosek S, Raftery D. Visible light driven V-doped TiO_2 photocatalyst and its photooxidation of ethanol. The Journal of Physical Chemistry B. 2001; **105**(14):2815-2819

[7] Lizama C, Freer J, Baeza J, Mansilla HD. Optimized photodegradation of Reactive Blue 19 on TiO_2 and ZnO suspensions. Catalysis Today. 2002;**76** (2-4):235-246

[8] Huang MH, Mao S, Feick H, Yan H, Wu Y, Kind H, et al. Room-temperature ultraviolet nanowire nanolasers. Science. 2001;**292**(5523):1897-1899

[9] Kong XY, Wang ZL. Spontaneous polarization-induced nanohelixes, nanosprings, and nanorings of piezoelectric nanobelts. Nano Letters. 2003;**3**(12):1625-1631

[10] Xie Y, He Y, Irwin PL, Jin T, Shi X. Antibacterial activity and mechanism of action of zinc oxide nanoparticles against *Campylobacter jejuni*. Applied and Environmental Microbiology. 2011; **77**(7):2325-2331

[11] Agus SG, Irmansyah I, Akhiruddin M. Dye-sensitized solar cell based on flower-like ZnO nanoparticles as photoanode and natural dye as photosensitizer. Журнал нано-та електронної фізики. 2016;**8**(2):02012-02011

[12] Madhusoodanan KN, Vimalkumar TV, Vijayakumar KP. Gas sensing application of nanocrystalline zinc oxide thin films prepared by spray pyrolysis. Bulletin of Materials Science. 2015;**38** (3):583-591

[13] Ong CB, Ng LY, Mohammad AW. A review of ZnO nanoparticles as solar photocatalysts: Synthesis, mechanisms and applications. Renewable and Sustainable Energy Reviews. 2018;**81**: 536-551

[14] Goswami DY. Advances in Solar Energy: An Annual Review of Research and Development. Vol. 10. Boulder: ASES; 1995. pp. 165-209

[15] Iqbal J, Jilani A, Hassan PZ, Rafique S, Jafer R, Alghamdi AA. ALD grown nanostructured ZnO thin films: Effect of substrate temperature on thickness and energy band gap. Journal of King Saud University-Science. 2016;**28**(4):347-354

[16] Woo HS, Kwak CH, Chung JH, Lee JH. Co-doped branched ZnO nanowires for ultraselective and sensitive detection of xylene. ACS Applied Materials & Interfaces. 2014;**6**(24):22553-22560

[17] Voon CH, Derman MNB, Hashim U, Foo KL. Effect of anodizing voltage on the formation of porous anodic alumina on Al-0.5 wt% Mn alloys. Advanced Materials Research. 2014;**925**:455-459

[18] Nemes D, Moldovan V, Bruj E, Jumate N, Vida-Simiti I. Porous anodic alumina films obtained by two step anodization. Bulletin of the Transilvania University of Brasov. Series I: Engineering Sciences. 2011;**4**(2):75

[19] Bhattacharyya B. Electrochemical Micromachining for Nanofabrication, MEMS and Nanotechnology. Massachusetts: William Andrew Applied Science Publishers, Imprint of Elsevier Inc.; 2015. p. 270

[20] Yousif AA, Habubi NF, Haidar AA. Nanostructure zinc oxide with cobalt dopant by PLD for gas sensor applications. Journal of Nano- and Electronic Physics. 2012;**4**(2):02007

[21] Benno G, Joachim K. Optical properties of thin semiconductor films; 2003

[22] Bakr NA, Salman SA, Shano AM. Effect of co doping on structural and optical properties of NiO thin films prepared by chemical spray pyrolysis method. International Letters of Chemistry, Physics and Astronomy. 2015;**41**:15-30

[23] Palik ED, editor. Handbook of Optical Constants of Solids. Vol. 3. Academic Press; 1998

[24] Yildirim OA, Arslan H, Sönmezoğlu S. Facile synthesis of cobalt-doped zinc oxide thin films for highly efficient visible light photocatalysts. Applied Surface Science. 2016;**390**:111-121

[25] Kao CY, Liao JD, Chang CW, Wang RY. Thermal diffusion of Co into sputtered ZnO:Co thin film for enhancing visible-light-induced photo-catalytic activity. Applied Surface Science. 2011;**258**(5):1813-1818

[26] Sutanto H, Wibowo S, Arifin M, Hidayanto E. Photocatalytic activity of cobalt-doped zinc oxide thin film prepared using the spray coating technique. Materials Research Express. 2017;**4**(7):076409

[27] Borhani Zarandi M, Amrollahi Bioki H. Effects of cobalt doping on optical properties of ZnO thin films deposited by sol–gel spin coating technique. Journal of Optoelectronical Nanostructures. 2017;**2**(4):33-44

[28] Gajbhiye SB. Photocatalytic degradation study of methylene blue solutions and its application to dye industry effluent. International Journal of Modern Engineering Research. 2012;**2**(3):1204-1208

Section 3

Nanoparticles

Chapter 4

Diverse Synthesis and Characterization Techniques of Nanoparticles

Agnes Chinecherem Nkele and Fabian I. Ezema

Abstract

Nanoparticles are small particles that range from 1 to 100 nm in size, exhibit several physical and chemical features. An understanding of nanoparticles would reveal great qualities and potential applications that would aid the diversification of thin film technologies. The synthesis methods employed like top-down, bottom-up, chemical, biological, and mechanical processes have great influence on the properties exhibited by such nanomaterials. This review covers an insight into the knowledge of nanoparticles, their classifications, parameters affecting their efficient performance, synthesis and characterization techniques of nanoparticles. Nanoparticles are also characterized to obtain their morphological, structural, optical, elemental, size, and physiochemical features. The potential applications of nanoparticles have not been left undiscussed.

Keywords: nanoparticles, thin film technologies, synthesis, characterizations, applications

1. Introduction

Nanotechnology involves synthesizing and developing different nanomaterials. The field of nanotechnology allows different nanoparticles of unique features to be produced. Nanoparticles (NPs) are complex material particles that fall within the range of one to hundred nanometers. Their nanometer sizes drive the chemical, optical, physical, and electric features of the nanoparticles [1]. Naturally, nanoparticles can be sourced from geological, biological, meteorological, and cosmological means. However, nanoparticles can be created from liquid and solid materials by breaking down biopolymers, condensing gases, wet chemical process, implantation of ions, hydrothermal process, pyrolysis, radiolysis etc. Nanoparticles are usually viewed with the aid of electron microscopes, can penetrate filters, and have unique mechanical properties that distinguish them from the bulk materials. Nanoparticles exist in various shapes like nanorods, nanostars, nanofibers, nanospheres, nanoflowers, nanoboxes etc. [2].

Nanoparticles comprise a functionalized surface, a shell of different layered materials, and the core/main nanoparticle [3]. The features of materials in their bulk form are different from their nanoparticle forms because of the large area to volume ratio, interfacial layer, affinity to solvents, kind of coating, quantum mechanics effects, rate of diffusion, mechanical, and ferromagnetic features [1]. The large area to volume ratio makes the nanoparticles highly reactive and able to

IntechOpen

penetrate membranes. The chemical nature of nanoparticles should be studied to enhance their molecular attachment to surfaces.

2. Classification of nanoparticles

Nanoparticles may be metallic, non-metallic [1], anthropogenic, engineered, organic, or inorganic as outlined in **Figure 1**. Metallic nanoparticles include copper,

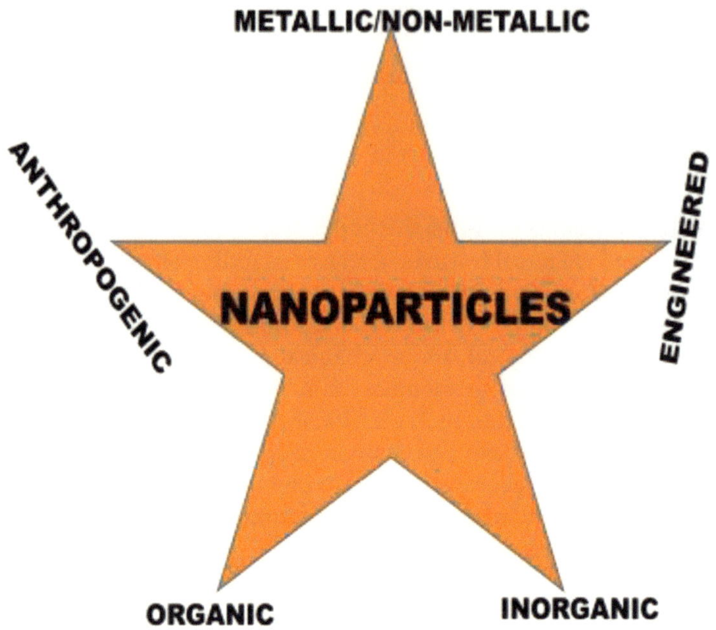

Figure 1.
Schematics on the classifications of nanoparticles.

Nanoparticles	Features
Silver	Very effective, high antimicrobial performance, wide range of usage
Gold	Good for identifying protein interactions, useful in tracing out fingerprints, detects antibiotics and cancerous cells, efficient for cancer diagnosis and other bacteria
Iron	Biocompatible and useful for treating cancer, sorting stem cells, analyzing genes, and drug delivery
Quantum dot	Diameters less than 10 nm, semiconducting nanoparticle, size-dependent
Carbon nanotubes	sp2 hybridized carbon atoms, strong electron bonds, high electrical conductivity, good catalysts.
Copper	Wide absorption spectrum, distinct optical features, yields good quality nanoparticles
Ceramics	Inorganic amorphous solids; could be polycrystalline, porous, amorphous or dense; vastly applied in photocatalysis, imaging devices etc.
Semiconductor	Large and tunable band gap nature, suitable in water splitting and electronic appliances.
Polymeric	Majorly organic components and easily functionalized
Lipid-based	Comprise lipid components, uses surfactants as core stabilizers

Table 1.
Some nanoparticles and their respective features.

magnesium, zinc, gold, titanium, silver etc.; while non-metallic nanoparticles include silica, carbon nanotubes etc. Anthropogenic NPs are by-products obtained from industrial produce while engineered nanoparticles are directly obtained from manufacturing processes.

Some of the nanoparticles and their features [2, 4] have been summarized in **Table 1**.

3. Synthesis techniques of nanoparticles

The techniques applied in synthesizing nanoparticles greatly influence their morphology, size, structure, and performance. The electrochemical, physiochemical, optical, and electrical features of the nanoparticles are also affected. In some occasions, nanoparticles are coated so as to retain their features after precipitating out of suspensions. The synthesis methods for nanoparticles are broadly divided into top-down and bottom-up approaches [4].

3.1 Top-down approach

Top-down method is a destructive method that breaks down large molecules into smaller parts before converting into the relevant nanoparticles. This approach involves some decomposition strategies like chemical vapor deposition (CVD), milling process, and physical vapor deposition (PVD). Milling is used to extract nanoparticles from coconut shells with the crystallite size reducing with increasing time. Nanoparticles of iron oxide, carbon, dichalcogenides, cobalt (III) oxide have been produced using this method.

3.2 Bottom-up approach

This approach involves the formation of nanoparticles from simple materials in a build-up manner. It is environmentally friendly, less poisonous, feasible, and of low cost. The materials used are usually Reduction and sedimentation processes like green synthesis, bio-chemical, spin coating, sol–gel etc. adopt this approach. Nanoparticles of titanium dioxide, gold, bismuth have been synthesized via this approach. The reaction chain for the production of gold nanoparticle has been illustrated in **Figure 2** [5].

Synthesizing nanoparticles could also involve chemical or biological processes [1]. Some chemical synthesis techniques of nanoparticles include sol–gel method, wet chemical synthesis, hydrothermal method, thermal decomposition, microwave method etc. [2]; while the biological means involve enzymes, microorganisms, plant extracts, and fungi.

3.3 Chemical methods

Some chemical methods adopted in synthesizing nanoparticles include sol gel, precipitation, hydrothermal, thermal decomposition, solvothermal, vapor synthesis etc. [6, 7]. Sol–gel method is an easy means of producing nanostructures by homogenously mixing precursors in a solvent to form a gel material which is then heated to produce the required nanoparticle. It begins from preparing a sol which undergoes gelation process to solvent removal. Wet chemical/precipitation method is a fast and easy process for synthesizing large scale nanoparticles. Hydrothermal method utilizes high pressure and temperature to power heterogeneous reactions under aqueous solvents like water. The kind of pressure, pH, and temperature

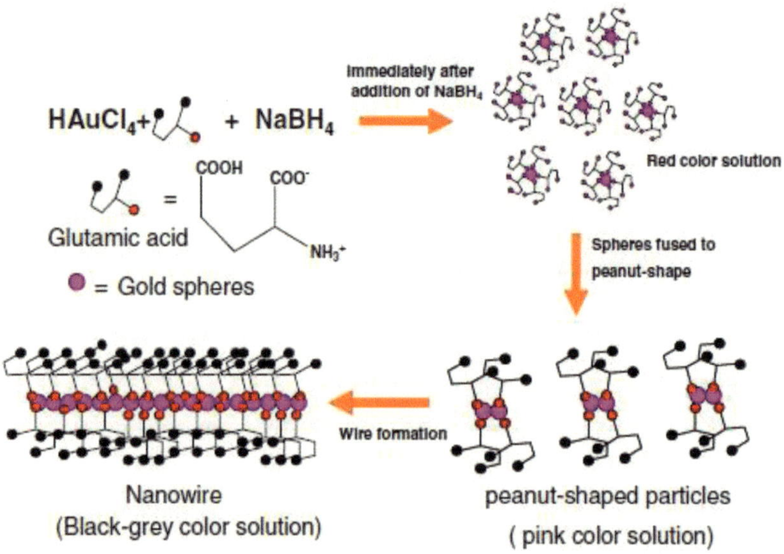

Figure 2.
Formation process of gold nanoparticle.

applied affects the features of the synthesized nanoparticles. Such nanoparticles are suitable for biotechnological use because of their hydrophilic surface nature [8]. Thermal decomposition involves oxidizing a solid material in optimal temperature. Solvothermal method uses a solvent to produce various materials like polymers, semiconductors, or metals at moderate or high pressure [9]. It produces novel and stable nanoparticles with controlled thicknesses and temperature. To synthesize nanodots; the cationic source is dissolved in suitable solvent alongside a surfactant which stabilizes the growth rate. Cadmium selenide, zinc oxide, zinc selenide are producible using this method and can be applied in magnetic and biotech industries [10]. In vapor synthesis, gaseous molecules chemically react to produce a phase which condenses and leads to particle growth. The higher the temperature, the faster the particles are formed. Different means of inducing homogenous nucleation include condensing inert gases, vaporizing a supersaturated material using a pulsed laser, generating a spark discharge by charging electrodes, sputtering the material with unreactive gaseous ions; or through some chemical methods like chemical vapor deposition, photothermal method, flame synthesis, or spray pyrolysis [11]. This method suitably yields nanoparticles of titania, carbon, and silica. Flame synthesis is commonly used to commercially produce silica, carbon black, optical fiber, and titania [12]. Particles produced by converting gases in furnace reactors or hot walls are usually very pure, although it produces agglomerated particles.

3.4 Biological methods

Biological or biosynthesis of nanoparticles is an environmentally-friendly, green, and non-toxic method involving microorganisms [13–15]. Nanoparticles of iron oxide, silver, nickel oxide, copper oxide, zinc ferrite have been synthesized using this method [16–22]. The location of the nanoparticle determines the point of synthesis; whether intracellular or extracellular [1]. Intracellular production of nanoparticles uses enzymes to move ions into the cells of microbes and produces smaller sized nanoparticles in the organism. Extracellular synthesis does not involve

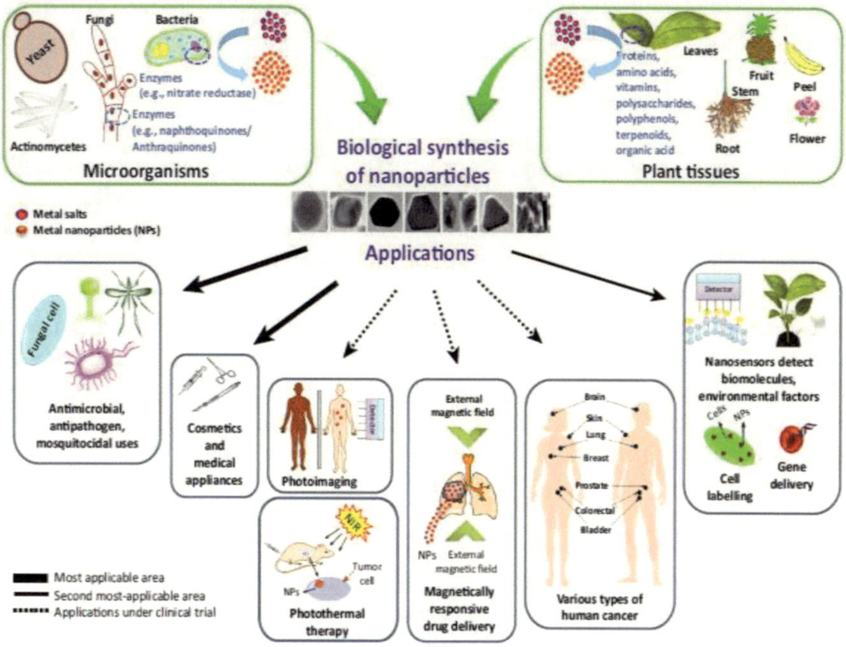

Figure 3.
Diverse bio-development synthesis of nanoparticles and their application areas.

cell components and yields nanoparticles outside the cell, uses fungi with large secretory organs. Microbes like fungi and bacteria are responsible for controlling the synthesis process. Microorganisms are immensely used to produce nanoparticles because of their economical, non-poisonous nature, and detoxification of heavy metal power. Phytonanotechnology is compatible with biological systems, available source materials, high stability, and entails synthesizing nanoparticles from plants [23]. Changes in the pH level of plants alter their binding strength, morphology, and the number of metallic ions available during the synthesis. The different sources, synthesis methods, and areas of application of nanoparticles have been represented in **Figure 3** [23]. Biogenic means of producing nanoparticles are green and cheap; with the involvement of fungi, waste materials, and bacteria [5].

3.5 Mechanical methods

Nanoparticles can also be synthesized by mechanical methods like mechanical alloying, milling, and mechanochemical processes [24]. Milling method regenerates interfacial chemical operations at low temperatures. Mechanochemical technique involves continuous welding operations that adequately select milling materials and minimize agglomerations. For effective production; the stoichiometry of source materials, thermal treatment, paths for reaction to occur, and milling conditions would be carefully considered. Nanoparticles of oxides, iron, nickel, silver, cobalt can be synthesized using these methods.

4. Characterization methods for nanoparticles

Properties of nanoparticles like shape, size, surface morphology, crystalline nature, light absorption etc. need to be completely described using relevant characterization techniques [2]. Some of the methods used to characterize nanoparticles [4] include:

4.1 Morphological features

The morphology of nanoparticles greatly influence the properties exhibited by nanoparticles. Microscopy methods applied on nanoparticles are usually electron microscopy or scanning probe microscopy. Scanning electron microscope (SEM) gives nanoscale and surface information of the dispersion and morphology of nanoparticles. Microscopy techniques are destructive and used for single-particle measurements. Transmission electron microscopy (TEM) uses transmittance of electrons to provide bulk information at high and low magnifications. Optical microscopic technique is not useful for nanoparticles because the size of nanoparticles is smaller than light diffraction limit. Coupling spectroscopic techniques to electron microscopes would enable elemental studies to be carried out.

4.2 Optical studies

Optical methods reveal reflectance, transmittance, photochemical, and luminescence features of nanoparticles. Spectroscopy uses the interaction of particles with electromagnetic radiation to determine the shape, concentration, and size of nanoparticles. Spectroscopic techniques like infrared, ultraviolet–visible, photoluminescence (PL), UV/vis-diffuse reflectance spectrometer (DRS), and magnetic resonance methods are applied to nanoparticles. DRS is specially used to determine the band gap energy of nanoparticles. PL studies reveal the effect of emissivity and absorptivity on the excitation of photons, half-life, and recombining effects of the charges. The sizes of nanoparticles affect their optical features and make it useful in bioimaging devices [4].

4.3 Structural analysis

The structure of nanoparticles gives details about the kind of bond existing between the atoms and the features of the bulk material. Some of the structural techniques used on nanoparticles include BET, X-ray diffractometry (XRD), IR etc. XRD describes the phase, particle size, type of NP, and crystal nature of the nanoparticles.

4.4 Elemental studies

The elemental composition of nanoparticles can be determined using energy dispersive X-ray spectroscopy (EDX), XPS, Raman, FT-IR etc. EDX details the elemental components of bulk particles. Better contrast is obtainable when the obtained spectra are compared with a computer generated model. XPS is a very sensitive spectroscopic method used to obtain the exact compositional ratio of the elements, their bonding nature, depth profile analysis. Raman and FTIR techniques use vibrational methods to show functionalized peaks and particle information.

4.5 Size estimation

Sizes of nanoparticles can be estimated using scanning electron microscope, transmission electron microscope, X-ray diffractometer, atomic force microscope etc. The sizes of the nanoparticles are obtained using size distribution profiles and give more precise results when used alongside digital models. The surface area can be estimated using BET via adsorption and desorption processes.

4.6 Physiochemical characteristics

Mechanical properties, optical activity, surface area, and chemical reactions of nanoparticles are physiochemical characteristics obtainable from nanoparticles. Free surface electrons on nanoparticles are very mobile and are not scattered upon light illumination. The magnetic features of NPs are manifested at small nanoscales due to their uneven distribution, influenced by the synthesis technique adopted, and find vast application in biomedicine, resonance imaging, and catalytic devices. Mechanical characteristics of nanoparticles like stress, surface coatings, hardness, strain, friction, adhesiveness etc. aid an understanding of NPs and greatly affect the quality of the surface. Nanoparticles have great conduction to heat especially on the surface.

5. Application areas of nanoparticles

Generally, nanoparticles have been applied in various areas including anticancer drugs, vaccines, disease treatment, cancer diagnosis, mechanical factories, electronics, optical devices, energy harvesters, manufacturing processes, cell imaging, and delivery systems due to their unique features [4]. NPs also aid water contaminants to be absorbed on the surface during water purification, serve as environmental sensors, and protect materials from harmful substances. Some of the application areas of the nanoparticles [2, 23] have been summarized in **Table 2**.

Despite the numerous applications of NPs; they suffer from poisonous and harmful body effects which inhaled, ingested, or transferred to the ground and surrounding environs. Nanoparticles are also affected by organic materials which lead to agglomeration. The poisonous effects associated with NP synthesis can be curtailed by adopting green synthesis methods especially in the synthesis of silver,

Nanoparticles	Application areas
Nickel oxide	Dye sensitized solar cells, supercapacitors, batteries, water treatment and catalytic systems, gas sensing devices.
Carbon nanotubes	Integrated circuits, electronic components, textile, construction, cosmetics, medicine
Cerium oxide	Biomedical equipments, electronic appliances, energy devices
Titanium dioxide	Coatings, water purifiers, paints
Silver	Clothing, textile industries, food packaging companies, agriculture, automotives, electronics, medicine, and fitness centers.
Iron	Optical devices, water purifier
Calcium	Agriculture, automotives, food
Zinc oxide	Agriculture, automotives, cosmetics, home appliances, food
Gold	Cosmetics, environmental products, food, medicine
Palladium	Automotive, electronic appliances, food

Table 2.
Application areas of some nanoparticles.

iron, copper, gold nanoparticles amongst others [25]. The synthesis process for silver nanoparticles is as shown in **Figure 4** [25].

Green synthesis involves different capping substances like biomolecules and polysaccharides. Green methods are non-poisonous, environmentally friendly, involve toxic-free solvents, compatible in biological systems, and utilize reagents like sugars, polymers, vitamins, plant extracts [26]. Plant-based extracts like latex, leaf, seed, root, or stem are more suited for bioprocesses as they are cheap, non-complex, easily reproduced, and highly stable. Other sources of waste materials useful for nanoparticle production have been outlined in **Figure 5**. Models can be developed to minimize the difficulties associated with distributing the size of the particles and NPs synthesis by computing the rates at which the particles get nucleated [11].

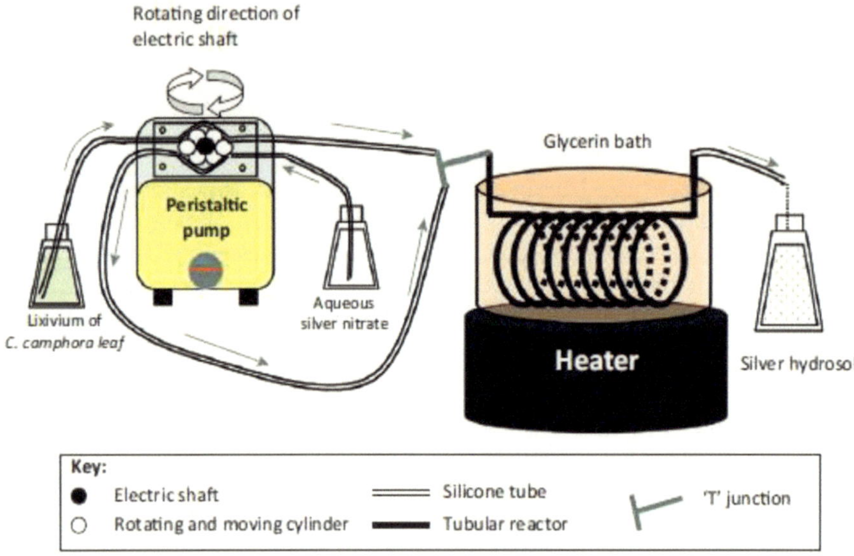

Figure 4.
Experimental diagram for the bioproduction of silver nanoparticles.

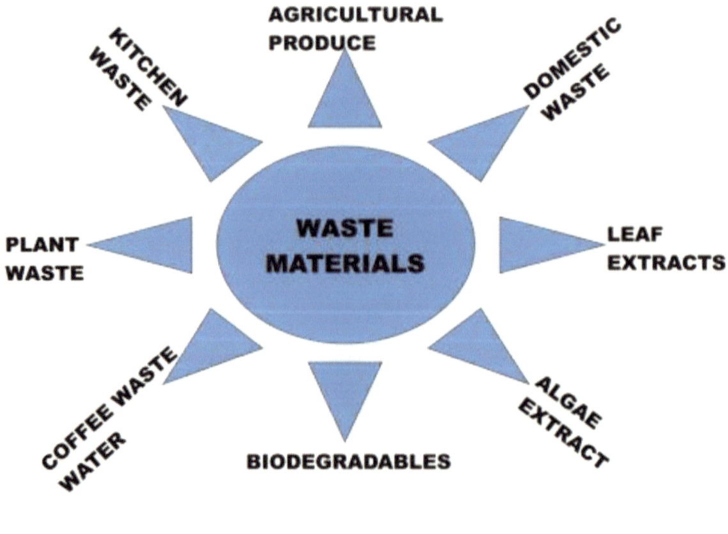

Figure 5.
Schematic showing different sources of waste material.

6. Conclusion and future perspective

The need for environmentally-friendly and stable nanomaterial that would be compatible with biological systems have prompted researchers into the production of nanoparticles. This chapter gives general knowledge on nanoparticles, their classification, merits and demerits, several synthesis and characterization techniques. Nanoparticles have economical and simple manufacturing processes that are classified into top-down method, bottom-up approach, chemical synthesis, biological method, and mechanical process. Several characterization methods of nanoparticles are geared towards understanding the morphological, structural, optical, size, mechanical, and physiochemical features. Each property is obtainable from different machines and using different techniques. The synthesis and characterization methods employed greatly influence the obtained features of the nanoparticles. Nanoparticles find useful application in medicine, drug delivery, cosmetics, optical devices, electronics, solar cell devices etc.

Author details

Agnes Chinecherem Nkele[1*] and Fabian I. Ezema[1,2,3,4]

1 Department of Physics and Astronomy, University of Nigeria, Nsukka, Enugu, Nigeria

2 Nanosciences African Network (NANOAFNET), iThemba LABS-National Research Foundation, 1 Old Faure Road, Somerset West 7129, P.O. Box 722, Somerset West, Western Cape Province, South Africa

3 UNESCO-UNISA Africa Chair in Nanosciences/Nanotechnology, College of Graduate Studies, University of South Africa (UNISA), Muckleneuk Ridge, P.O. Box 392, Pretoria, South Africa

4 Department of Physics, Faculty of Natural and Applied Sciences, Coal City University, Enugu, Nigeria

*Address all correspondence to: nechenkele@gmail.com

References

[1] S. Hasan, "A Review on Nanoparticles: Their Synthesis and Types," vol. 4, p. 3, 2015.

[2] "Nanoparticle," *Wikipedia*. Aug. 06, 2020, Accessed: Aug. 15, 2020. [online]. Available: https://en.wikipedia.org/w/index.php?title=Nanoparticle&oldid=971565062.

[3] Christian P, Von der Kammer F, Baalousha M, Hofmann T. Nanoparticles: Structure, properties, preparation and behaviour in environmental media. Ecotoxicology. Jul. 2008;**17**(5):326-343. DOI: 10.1007/s10646-008-0213-1

[4] Khan I, Saeed K, Khan I. Nanoparticles: Properties, applications and toxicities. Arabian Journal of Chemistry. 2019;**12**(7):908-931

[5] Sharma D, Kanchi S, Bisetty K. Biogenic synthesis of nanoparticles: A review. Arabian Journal of Chemistry. Dec. 2019;**12**(8):3576-3600. DOI: 10.1016/j.arabjc.2015.11.002

[6] Saranya M, Santhosh C, Ramachandran R, Nirmala Grace A. Growth of CuS nanostructures by hydrothermal route and its optical properties. Journal of Nanotechnology. 2014;**2014**

[7] Kaviyarasu K, Manikandan E, Paulraj P, Mohamed SB, Kennedy J. One dimensional well-aligned CdO nanocrystal by solvothermal method. Journal of Alloys and Compounds. 2014;**593**:67-70

[8] V. K. Yadav et al., "Microbial synthesis of nanoparticles and their Applications for wastewater treatment," in Microbial Biotechnology: Basic Research and Applications, J. Singh, A. Vyas, S. Wang, and R. Prasad, Eds. Singapore: Springer, 2020, pp. 147-187.

[9] Y. F. Liu, G. H. Yuan, Z. H. Jiang, and Z. P. Yao, "Solvothermal Synthesis of Mn3O4 Nanoparticle/Graphene Sheet Composites and Their Supercapacitive Properties," *J. Nanomaterials*, vol. 2014, pp. 151:151-151:151, Jan. 2014, doi: 10.1155/2014/190529.

[10] Gersten B. Solvothermal synthesis of nanoparticles. Chem. 2005;**5**:11-12

[11] Swihart MT. Vapor-phase synthesis of nanoparticles. Current Opinion in Colloid & Interface Science. Mar. 2003;**8**(1):127-133. DOI: 10.1016/S1359-0294(03)00007-4

[12] Kammler HK, Mädler L, Pratsinis SE. Flame synthesis of nanoparticles. Chemical Engineering & Technology. 2001;**24**(6):583-596. DOI: 10.1002/1521-4125(200106)24:6<583::AID-CEAT583>3.0.CO;2-H

[13] Aisida SO et al. Incubation period induced biogenic synthesis of PEG enhanced Moringa oleifera silver nanocapsules and its antibacterial activity. Journal of Polymer Research. 2019;**26**(9):225

[14] Aisida SO, Akpa PA, Ahmad I, Maaza M, Ezema FI. Influence of PVA, PVP and PEG doping on the optical, structural, morphological and magnetic properties of zinc ferrite nanoparticles produced by thermal method. Physica B: Condensed Matter. 2019;**571**:130-136

[15] Aisida SO, Akpa PA, Ahmad I, Zhao T, Maaza M, Ezema FI. Bio-inspired encapsulation and functionalization of iron oxide nanoparticles for biomedical applications. European Polymer Journal. 2020;**122**:109371

[16] Madubuonu N et al. Biosynthesis of iron oxide nanoparticles via a composite of Psidium guavaja-Moringa

oleifera and their antibacterial and photocatalytic study. Journal of Photochemistry and Photobiology B: Biology. 2019;**199**:111601

[17] Aisida SO et al. Biogenic synthesis and antibacterial activity of controlled silver nanoparticles using an extract of Gongronema Latifolium. Materials Chemistry and Physics. 2019;**237**:121859

[18] Aisida SO et al. Biogenic synthesis of iron oxide nanorods using Moringa oleifera leaf extract for antibacterial applications. Applied Nanoscience. 2020;**10**(1):305-315

[19] Ugwoke E et al. Concentration induced properties of silver nanoparticles and their antibacterial study. Surfaces and Interfaces. 2020;**18**:100419

[20] Nwanya AC et al. Industrial textile effluent treatment and antibacterial effectiveness of Zea mays L. dry husk mediated bio-synthesized copper oxide nanoparticles. Journal of Hazardous Materials. 2019;**375**:281-289

[21] Nwanya AC et al. Maize (*Zea mays* L.) fresh husk mediated biosynthesis of copper oxides: Potentials for pseudo capacitive energy storage. Electrochimica Acta. 2019;**301**:436-448

[22] Nwanya AC et al. *Zea mays* lea silk extract mediated synthesis of nickel oxide nanoparticles as positive electrode material for asymmetric supercabattery. Journal of Alloys and Compounds. 2020;**822**:153581

[23] Singh P, Kim Y-J, Zhang D, Yang D-C. Biological synthesis of nanoparticles from plants and microorganisms. Trends in Biotechnology. Jul. 2016;**34**(7):588-599. DOI: 10.1016/j.tibtech.2016.02.006

[24] Tsuzuki T, McCormick PG. Mechanochemical synthesis of nanoparticles. Journal of Materials Science. Aug. 2004;**39**(16):5143-5146. DOI: 10.1023/B:JMSC.0000039199.56155.f9

[25] Kharissova OV, Dias HVR, Kharisov BI, Pérez BO, Pérez VMJ. The greener synthesis of nanoparticles. Trends in Biotechnology. Apr. 2013;**31**(4):240-248. DOI: 10.1016/j.tibtech.2013.01.003

[26] Gour A, Jain NK. Advances in green synthesis of nanoparticles. Artificial Cells, Nanomedicine, and Biotechnology. Dec. 2019;**47**(1):844-851. DOI: 10.1080/21691401.2019.1577878

Chapter 5

Roles of Cobalt Doping on Structural and Optical of ZnO Thin Films by Ultrasonic Spray Pyrolysis

Sabrina Roguai and Abdelkader Djelloul

Abstract

Here we report a systematic study of structural, optical, and magnetic measurements of $Zn_{1-x}Co_xO$ (x = 0–0.22 at.%) by Ultrasound pyrolysis spray technique. The hexagonal wurtzite structure of our films is confirmed by X-ray diffraction with an average crystallite size estimated in the range of 18–30 nm. For the optical proprieties, using the Levenberg–Marquardt least squares rule, the experimental transmission measurements were perfectly adapted to the transmission data calculated by a combination of the Wemple-DiDomenico model, the absorption coefficient of an electronic transition and the Tauc-Urbach model. The concentration of the NCo absorption centers and the oscillator intensity f of the d-d transition of Co2+ ions are determined by the Smakula method. The presence of high concentrations of localized states in the thin films is responsible for the reduction in the width of optical bandgap.

Keywords: thin films, X-ray diffraction, optical properties

1. Introduction

Zinc oxide is a transparent direct gap and wide band gap semiconductor (3.37 eV) with a fairly high excitonic binding energy of 60 meV at room temperature, it is attracting more and more attention from researchers because of its wide range applications, in particular in the field of spintronics [1–4]. Zinc oxide is a transparent material with a transmission value of 0.9 in the visible, crystallizes in a wurtzite-like structure defined by a hexagonal lattice where zinc ions occupy the centers of the tetrahedral sites and oxygen ions occupy the vertices.

The improvement of the properties of thin ZnO layers is commonly achieved through doping. Among the different dopants, cobalt (Co) thanks to the similarity between the ionic rays (r_{Co2+} = 0.058 nm) and (r_{Zn2+} = 0.060 nm). In order to obtain what is called Diluted Magnetic Semi-conductor (DMS), these DMS play an important role as they allow the integration of certain components of spintronics and optoelectronics [5–10].

Different technological processes can be used to deposit cobalt-doped ZnO in thin films [11–14]. The doping of ZnO with transition metal ions such as Co (substitution of Zn + 2 ions by Co + 2 ions) induces magnetic properties due to its possible applications in the field of spintronics. Moreover, the excellent optical properties of ZnO and the engineering possibility of the bandgap by doping the matrix with Co2+ ions strongly encourages the exploration of the magneto-optical properties of the Co2+ doped ZnO system [15–17]. Much research work is focused on the elaboration of single-phase Zn1-xCoxO thin films with several techniques [18].

The determination of the optical constants (refractive index, extinction coefficient, thickness) of thin films is of great interest both on a fundamental and technological level. Among the various techniques commonly used to deduce these parameters, we can cite the techniques of optical transmission, X-ray reflectometry and ellipsometry.

ZnO thin films, depending on the cobalt doping concentration, have a high transparency of more than 90% in the visible (400–750 nm) and infrared range. Three absorption bands located at 568, 608 and 659 nm that can be attributed to the dd transitions of the tetrahedrally coordinated Co2+ ion in the high spin state are attributed respectively: A_2 4 (F) → A_1 (G) $_2$, A_2 4 (F) → T_1 (P) $_4$ and A_2 4 (F) → E (G) $_2$. The value of *Eg* for the ZnO thin film was found to be increased from 3.26 eV to 3.31 eV with an increase in the Co doping concentration. The increase of the bandgap can be attributed to the strong sp-d exchange interaction between the band electrons and the localized d-electrons of the dopant [19].

In this study, thin layers of ZnO, $Zn_{1-x}Co_xO$ deposited by the ultason pyrolysis spray technique at a temperature of 450 °C weighing 30 min.. and their structural and optical properties are reported in this work. A particular attention is given to the theoretical methods used for the determination of the dispersion parameters of the films using only a single transmission spectrum.

2. Experimental

2.1 Film preparation

ZnO, $Zn_{1-x}Co_xO$ thin films have been developed by the pyrolysis spray technique, is well known for its simplicity and possibility to produce large area films. The properties of the deposited material can be changed and controlled by appropriate optimization of the deposition conditions, on glass substrates (solid glass), the choice of glass as the deposition substrate was adopted because of the good thermal expansion it presents with ZnO (α_{verre} = $8.5.10^{-6}$ K^{-1}, α_{ZnO} = $7.2.10^{-6}$ K^{-1}) so as to minimize the stresses at the substrate layer interface, and for economic reasons, for their transparency which is well suited for the optical characterization of films in the visible.

We used, in this work, 0.01 M of zinc acetate [$Zn(CH_3COO)_2.2H_2O$)] (Fulka 99.9%);comme matériau source (de ZnO) que nous avons dissous dans 50 ml deionized water (resistivity = 18.2 MΩ.cm); 20 ml CH_3OH (Merck 99.5%); 30 ml C_2H_5OH (Merck 99.5%) and Cobalt nitrate hexahydrate 1–22% (Co, at. %) [Co $(NO_3)_2{\cdot}6H_2O$] has been used as the Co source. A small amount of acetic acid was added to the aqueous solution for adjusting the pH value to about 4.8, in order to prevent the formation of hydroxides. Further details are reported elsewhere [20].

The structural properties of the thin films were studied by an X-ray diffractometer Rigaku Ultima IV powder equipped with CuKα radiation (λ = 1.54 Å). X-ray diffraction (XRD) data were recorded at a scanning speed of 2 degree/min, between 20° and 80°. The optical properties of the films were studied by recording the transmittance spectra of the films within the wavelength range of 190–1800 nm using a Perkin Elmer UV–VIS–NIR Lambda 19 spectrophotometer.

3. Results and discussion

3.1 Structure and microstructure analysis

Figure 1 shows the DRX diagram of a thin layer of cobalt doped ZnO. It can be seen from the spectrum that the films are well crystallized. The different diffraction peaks correspond to the (100), (002), (101), (102), (110), (103), (112) planes of hexagonal wurtzite structure of polycrystalline ZnO (Card. JCPDS N0 36–1451), and no other crystalline phase was detected. Peak intensities vary considerably depending on the Co. The preferred direction of growth during deposition changes with the level of Co doping: along (002) for pure ZnO, (101) for xCo = 0.01, then again (002) for xCo in the range of 0.03–0.14, and finally at varying orientations for even higher Co content.

Table 1 illustrates the various results of X-ray diffraction Rietveld refinements results. It can be noticed that the size of crystallite varies within a narrow interval of 18–30 nm, whereas the micro-strain changes considerably with Co content in to the range of 0.131–0.289%.

The evolution of the mesh parameter a and c of the Zn1-xCoxO thin films as a function of the doping rate is shown in **Table 1**. Indeed, the mesh parameters were calculated using the Bragg's law and from the peak (002) of the X-ray diffractogram. An increase of the a mesh parameter is noted while the c parameter was not affected, accompanied with an increase of the unit volume when the cobalt level increases. However, the development of network parameters with a Co rate is more difficult than expected. Four points must be taken as a reference, as their effects occur concomitantly:

1. According to Vegard's law, the lattice parameters are expected to decrease slightly as the Co level increases, the ionic radius of Co is smaller than that of Zn. In fact, in our case, the ionic radius of Co is slightly smaller than that of Zn^{2+} in tetrahedral coordination with an ionic radius of 0.060 nm will not be completely filled by Co^{2+} in tetrahedral coordination with an (r_{Co2+} = 0.058 nm) but also with Co^{2+} in octahedral coordination with an r = 0.065 nm (low spin) or/and 0.075 nm (high spin) [21]. The substitution of Co into tetrahedral coordination was confirmed by transmittance measurements (see section 3.2).

2. Depending on the size impact in the nanomaterials, the lattice contraction is assumed to occur when the particle size (crystallites) reduces [22].

3. The mismatch between the glass substrate and the thin film (ZnO) deposited causes a certain constraint in the ZnO and may cause some distortion of the lattice, so that the lattice parameters may vary depending on this [23].

4. Influence of the deposition temperature (450 °C) on the stoichiometry, which means that possible gaps in O and/or Zn are likely to be created, and that O can even fill interstitial sites [24]. This will lead to anisotropic changes in the properties of the matrix.

5. Several Zn or/and Co could also invest interstitial sites, which would cause a disturbance of the layout parameters of the lattice.

3.2 Optical properties

The refractive index (n) is very important in the determination of the optical properties of semiconductors. Knowledge of the refractive index is essential in the

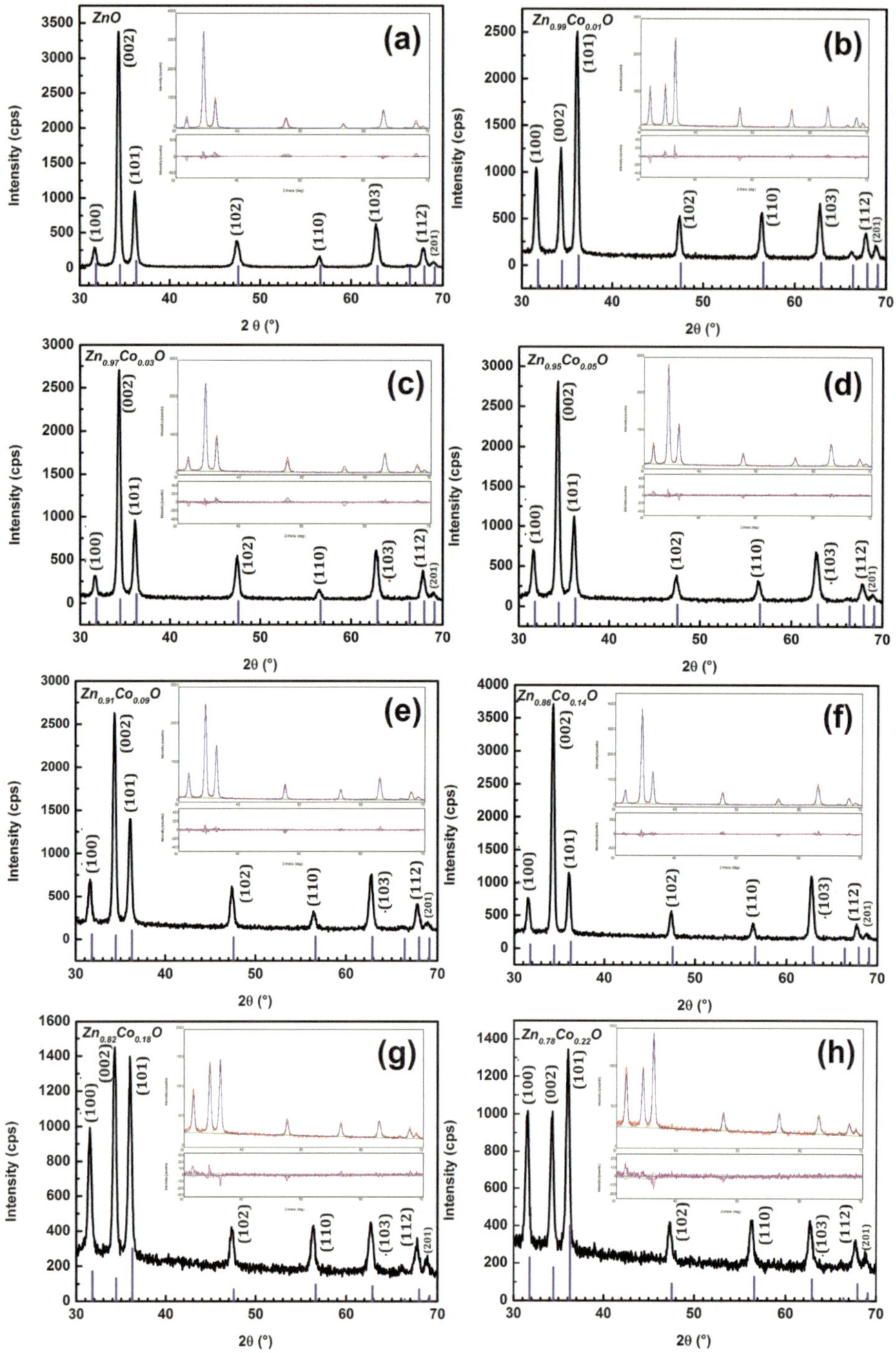

Figure 1.
XRD patterns of (a) pure ZnO film and (b–h) Co-doped $Zn_{1-x}Co_xO$ thin films. The XRD (b–h) correspond to the Co atomic content of 1%, 3%, 5%, 9%, 14%, 18% and 22%, respectively. The inset shows the Rietveld refinements of $Zn_{1-x}Co_xO$ composition. Solid blue curve: Calculated pattern; red solid line: Experimental data; solid orange line (down): Intensity difference.

design of laser heterostructures, optoelectronic devices, and solar cell applications. From the transmission spectra obtained for the ZnO, Zn1-xCoxO layers, the refractive index can be determined. The refractive index of the film can be calculated using a single-effective-oscillator fit, proposed by Wemple et al. [25], of the form

Composition	Crystallite Size (nm)	Microstrain (%)	Lattice Parameters (Å)	R_{wp}(%)	Refinements R_p (%)	Factors R_e (%)	S	χ^2
ZnO	25	0.266	*a* = 3.2542 *c* = 5.2129	16.87	11.84	8.93	1.71	2.92
$Zn_{0.99}Co_{0.01}O$	21	0.096	*a* = 3.2577 *c* = 5.2124	9.21	7.37	7.55	1.21	1.48
$Zn_{0.97}Co_{0.03}O$	35	0.230	*a* = 3.2594 *c* = 5.2172	8.23	6.28	5.40	1.52	2.32
$Zn_{0.95}Co_{0.05}O$	23	0.289	*a* = 3.2572 *c* = 5.2162	9.58	7.41	7.63	1.25	1.57
$Zn_{0.91}Co_{0.09}O$	22	0.131	*a* = 3.2613 *c* = 5.2172	8.55	6.64	6.76	1.26	1.59
$Zn_{0.86}Co_{0.14}O$	25	0.143	*a* = 3.2605 *c* = 5.2148	8.94	6.99	6.15	1.45	2.11
$Zn_{0.82}Co_{0.18}O$	18	0.147	*a* = 3.2689 *c* = 5.2230	8.40	6.72	6.18	1.36	1.85
$Zn_{0.78}Co_{0.22}O$	18	0.142	*a* = 3.2644 *c* = 5.2167	8.02	6.55	6.12	1.31	1.71

Table 1.
X-ray diffraction Rietveld refinements results.

$n^2 - 1 = E_d E_0 / (E_0^2 - E^2)$ where $E = hc/\lambda$ is the photon energy, E_0 the single oscillator energy, and E_d is the dispersion energy. The parameter E_d, which is a measure of the strength of inter-band optical transitions, is found to obey the simple empirical relationship $E_d = \beta N_C Z_a N_e$ in which NC is the coordination number of the cation closest to the anion, Za is the formal chemical valence of the anion, Ne is the effective number of valence electrons per anion (usually Ne = 8), and for the ionic βi = 0.26 ± 0.04 eV. The values of E0, Ed and β of ZnO are listed in **Table 2**.

Information on the dispersion of the refractive index below the band gap can be evaluated by the dispersion correlation below:

$$n^2 - 1 = \frac{S_0 \lambda_0^2}{1 - (\lambda_0/\lambda)^2} \tag{1}$$

in which λ is the wavelength of the incident light, S0 is the average oscillator power of the absorption band with the resonance wavelength λ0, which is an average oscillator wavelength. The equation Eq. (1) is also transformed into:

$$n^2 - 1 = \frac{(n_\infty^2 - 1)\,\lambda^2}{\lambda^2 - \lambda_0^2} \tag{2}$$

where n_∞ and λ_0 are the high-frequency refractive index and average oscillator wavelength, respectively.

When absorption bands in the visible and near infrared regions coexist (extinction coefficient, $k \neq 0$), the refractive index dispersion data can be analyzed by the following dispersion relation:

$$n^2 - 1 - k^2 = \frac{(n_\infty^2 - 1)\,\lambda^2}{\lambda^2 - \lambda_0^2} \tag{3}$$

Cristal	E_0 (eV)	E_d (eV)	M_{-1}	M_{-3}, 10^{-2} $(eV)^{-2}$	n_∞	*n* at 598 nm	β (eV)
ZnO	6.4	17.1	2.672	6.523	1.916	1.996	0.27

Table 2.
Dispersion parameters for ZnO [34]. Wurtzite (N_C = 4, Z_a = 2, N_e = 8).

In cases where the absorption of a chemical system shows the "simple form" band of absorption, the electronic transition is capable of correctly representing the same band. To replicate the structure of this absorption band, a simple Gaussian profile centred on the vertical transition in question is then used. This assumes a vertical electronic transition between the state Si and the state Sj, the wavelength o electron transition $\lambda_{i \to j}$ and the strength of the oscillator $f_{i \to j}$ The resulting spectral band's expression $\alpha_{i \to j}$ is proportional to a Gaussian function such as:

$$\alpha_{i \to j}(\lambda) \propto \frac{f_{i \to j}}{\xi' \sqrt{\pi}} \exp\left(-\frac{(\lambda - \lambda_{i \to j})^2}{\xi'^2}\right), \xi' = \frac{\xi}{2\sqrt{\ln(2)}} \tag{4}$$

where ξ represents the width at half maximum of the Gaussian function, or bandwidth. This parameter is chosen empirically by comparison with experiment.

In a simple solid consisting of a host lattice and an impurity ion, the absorption coefficient α for the solid solution can be considered as the sum $\alpha = \alpha_h + \alpha_i$, where α_h, is the absorption from the host lattice and α_i is the contribution to the absorption coefficient from the impurity ion.

Where ξ the distance at the half limit of the Gaussian function or bandwidth is represented. Compared to experiments, this parameter is selected empirically.

The absorption coefficient α for the solid solution can be considered as the sum $\alpha = \alpha_h + \alpha_i$, I where α_h, is the absorption from the host lattice and α_I is the contribution to the absorption coefficient from the impurity ion, in a simple solid consisting of a host lattice and an impurity ion.

α_h is equal to the absorption coefficient for the un-doped ZnO for ZnO:Co.

The extinction coefficient k is related by the expression $4\pi k/\lambda$ to the absorption coefficient α. In the transparent range ($\lambda \geq \lambda_g$), the extinction coefficient k is [29]:

$$k = k_0 \frac{(\exp(B\lambda_g/\lambda) - 1)}{(\exp(B) - 1)} + \frac{\lambda}{4\pi}\left[\alpha_0 + \sum_{j=1}^{q} \alpha_{i \to j}(\lambda)\right] \tag{5}$$

where λ_g - wavelength of absorption region ($E_g(eV) = 1239.8/\lambda_g(nm)$), i - ground state, j - excited state and q is the number of excited states.

The extinction coefficient k in the region of interband transitions ($\lambda \leq \lambda_g$) is:

$$k = k_1 \left(1 - \frac{\lambda}{\lambda_g}\right)^r + k_0 \tag{6}$$

In which k_0, k_1, B, λ_g, $f_{i \to j}$, ξ' and $\lambda_{i \to j}$ are the fitting parameters, r can, depending on the nature of the interband electronic transitions, have values such as 1/2, 3/2, 2, and 3, such as direct permitted, direct prohibited, indirect permitted and indirect prohibited transitions, respectively [26, 27]. The value of r for ZnO is always 1/2, i.e. the fundamental absorption corresponds to the direct transformation permitted.

Formulas relating the measured values of T(λ) and thickness, d, to the real and imaginary components of the refractive index, N = *n-ik*, for the absorbing film on a transparent substrate are necessary for the measurement of the optical constants from the data. The common approach is to consider light reflection and transmission at the three interfaces of the multilayer structure of the air/film/substrate/air and to express the results in terms of Fresnel coefficients.

The device is encircled by air with a refractive index of $n_0 = 1$. Taking into account all the multiple reflections at the three interfaces, it can be seen that the expression of the transmittance T(λ) for normal incidence is given in the case where $k^2 < < n^2$ is given in [20]:

$$T = \frac{A\chi}{B - C\chi + D\chi^2} \tag{7}$$

where,

$A = 16\gamma^2 n_s (n^2 + k^2)$

$B = \left[(n+1)^2 + k^2\right]\left[(n+1)\left(n + n_S^2\right) + k^2\right]$

$C = 2\eta\left[\left(n^2 - 1 + k^2\right)\left(n^2 - n_S^2 + k^2\right) - 2k^2\left(n_S^2 + 1\right)\right]\cos\varphi - 2k\eta\left[2\left(n^2 - n_S^2 + k^2\right)\right.$

$\left.+\left(n_S^2 + 1\right)\left(n^2 - 1 + k^2\right)\right]\sin\varphi D = \eta^2\left[(n-1)^2 + k^2\right]\left[(n-1)\left(n - n_S^2\right) + k^2\right]$

$\varphi = 4\pi n d/\lambda$

$\chi = \exp(-\alpha d)$

$\alpha = 4\pi k/\lambda$

$\gamma = \exp\left[-\frac{1}{2}(2\pi\sigma/\lambda)^2(1-n)^2\right], \eta = \exp\left[-2(2\pi\sigma/\lambda)^2\right].$

When σ is the height of surface irregularity for rms. The parameters n and k are the real and imaginary components of the refractive index of the thin film, d is the thickness of the film and ns is the refractive index of the real substratum. Knowing the substrate's refractive index and putting the n and k values as calculated from Eqs. (3), in Eqs. (5) and (6) respectively, in Eq. (7), it is possible to obtain the theoretical transmittance value, referred to as T_{Theo}. Then the experimental transmittance data (T_{expt}) was completely fitted with the transmittance data measured (T_{theo}) by Eq. (7) by the application of the Levenberg–Marquardt least square method, through a combination of the Wemple-DiDomenico model, the electronic transition absorption coefficient, and the Tauc-Urbach model.

The exact film thickness and energy bandgap can be determined by minimizing the sum of squares ($|T_{expt}$-$T_{theo}|$) created for different values of thickness (d) and gap wavelength (λg) by iterative technique and finding the corresponding n and k.

The glass substrate refractive index, taken from Ref. [28], is:

$$n_S^2 = 1 + \frac{1.0396 \times \lambda^2}{\lambda^2 - 6.0069 \times 10^3} + \frac{0.23179 \times \lambda^2}{\lambda^2 - 2.0017 \times 10^4} + \frac{1.0104 \times \lambda^2}{\lambda^2 - 1.0356 \times 10^8} \tag{8}$$

Transmittance spectra were taken at room temperature to study the optical properties of $Zn_{1-x}Co_xO$ films. At wavelengths of 565, 611, 657, 1297, 1410 and 1648 nm, the transmittance spectra of all films display the characteristic Co^{2+} absorption in the visible and near infrared spectral area. The prevalent absorption is the first three peaks.

The colorless host lattice (ZnO) is converted into green by the dopant (Co^{2+}) ionIf the concentration of dopant ions is low, it is possible to ignore the interaction between the dopant ions. This is what has been perceived as an isolated center of absorption here. In the studied systems, the actual distance between two Zn atoms is around ~0.326 nm, while the Zn atoms in $Zn_{1-x}Co_xO$ are present with ZnO distances of 0.196 nm in tetrahedral structures. The mean distance between Co ions that have been replaced by Zn sites within the ZnO crystal lattice can be calculated for uniformly distributed Co ions through [29] $N_{at} = (4/3)\ (Z/V_C)\ \pi\ r^3$, where r is the average radius of an atomic sphere and Nat is the number of atoms in the sphere. The structural parameters for $Zn_{0.95}Co_{0.05}O$ are: $a = 0.32572$ nm, $c = 0.52162$ nm; volume of unit cell $(V_C) = 47.92 \times 10^{-3}\ nm^3$; Z = 2. For zinc atoms in a wurtzite structure of $Zn_{0.95}Co_{0.05}O$, $Z/V_C = 41.73\ nm^{-3}$. Co ion occupy 5% of the zinc sites in the $Zn_{0.95}Co_{0.05}O$film. The 20th position of zinc from the probe atom is also occupied by cobalt. The average distance between Co ions is calculated to be about ~0.48 nm using the above measurement, indicating that the closest Co ion to a Co ion probe is located in the next unit cell.

According to the ligand field theory [20], splitting of $3d^7$ (Co^{2+}) orbital should result in the spectroscopic terms $^4A_2$ (*A*: no degeneracy), $^4T_2$, $^4T_1$ (*T*: three fold degeneracy), and ^2E (*E*: two fold degeneracy). For Co^{2+} in ZnO crystal lattice, Co^{2+} substitutes for some Zn^{2+}, and adopts tetrahedral ligand coordination. The 3d levels are extremely host sensitive. The strong crystal field in ZnO leads to the splitting of 3d electron orbits of Co^{2+} and produces the ground level: $^4A_2$, and the excited states: ^2E, $^4T_2$, and $^4T_1$, etc. The transitions from $^4A_2$ to $^4T_2$, and $^4T_1$ are spin-allowed.

Figures 2 and **3** show UV spectra of ZnO and $Zn_{1-x}Co_xO$ films. The absorption peaks located at 657, 610, and 567 nm for $Zn_{1-x}Co_xO$ films can be assigned as $^4A_2(F) \rightarrow {}^2E(G)$, $^4A_2(F) \rightarrow {}^4T_1(P)$, and $^4A_2(F) \rightarrow {}^2A_1\ (G)$, of Co^{2+}, attributed to the crystal field transitions in the high spin state of Co^{2+} in the tetrahedral coordination, suggesting that the tetrahedrally coordinated Co^{2+} ions substitute for Zn^{2+} ions in the hexagonal ZnO wurtzite structure [20]. Between 1272 and 1647 nm, additional crystal field transition was observed, namely $^4A_2(F) \rightarrow {}^4T_1(F)$ transition.

In **Figures 2** and **3**, the strong curve, refer to the fitting of the curve using Eq. (7) and the symbol represents the data from the experiments. The figures indicate a fair fit for the experimental results, indicating the precise determination of the parameters of Eq. (7). **Table 3** presents the values of d, Eg, Ed, E0, rms and n, derived by fitting the experimental data with Eq. (7).

The pure ZnO film optical energy band-gap was measured to be 3.26 eV. This value is marginally lower than the bulk value of 3.31 eV [1], and is in good alignment with the ZnO thin film data previously reported [30].

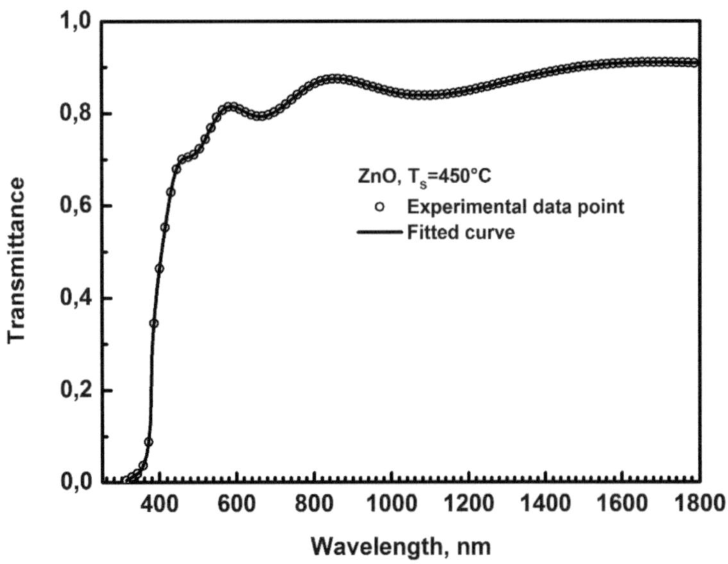

Figure 2.
Transmission spectrum of ZnO films deposited on glass substrate at 450 °C is presented as a reference. Measured (full circles) and calculated (solid lines) transmittance spectra of films.

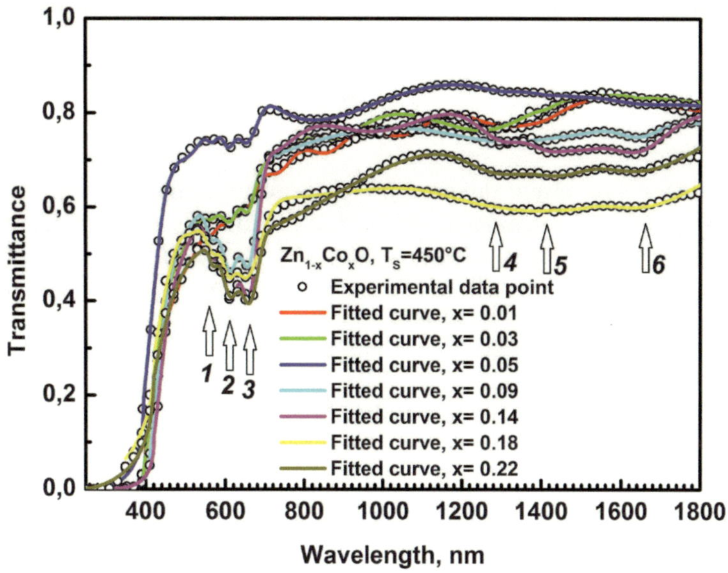

Figure 3.
Transmission spectra of $Zn_{1-x}Co_xO$ films deposited on glass substrate at 450 °C. measured (full circles) and calculated (solid lines) transmittance spectra of films.

The direct optical bandgap value will be reduced from 3.26 to 3.00 eV. The interactions of s-d and p-dexchange lead to a negative and positive correction of the conduction band and the edges of the valence band, resulting in the narrowing of the band gap. The interaction results in energy band corrections; the conduction band is lowered while the valence band is increased, causing the band gap to decrease [20].

The decrease in energy value from 3.26 eV (pure ZnO) to 3.00 eV ($Zn_{0.78}Co_{0.22}O$) appears to be due to active transitions involving 3d Co^{2+} ion levels and intense sp–d interactions between the itinerant 'sp' carriers (band electrons) and the dopant's localized 'd' electrons. Several researchers have already recorded this red change of band-gap Eg with the incorporation of Co^{2+} into ZnO [31].

Using single oscillator energy (E_O) and dispersion energy (E_d) obtained from the fitted transmittance spectra reported in **Table 3**, M_{-1} and M_{-3} moments of the optical spectra can be determined from the following two Equations [25]:

	Thickness, nm	E_g, eV	E_d, eV	E_0, eV	*n* at 598 nm	n_∞	M_{-1}	M_{-3}, $\times 10^{-2}$ $(eV)^{-2}$	σ, nm	Porosity %
ZnO	486	3.258	11.334	6.018	1.771	1.698	1.883	5.200	38	17.0
$Zn_{0.99}Co_{0.01}O$	1350	3.177	12.373	6.198	1.802	1.731	1.996	5.196	128	14.1
$Zn_{0.97}Co_{0.03}O$	846	3.097	13.478	6.185	1.859	1.783	2.179	5.696	130	09.7
$Zn_{0.95}Co_{0.05}O$	421	3.050	13.422	5.635	1.940	1.839	2.382	7.501	95	03.8
$Zn_{0.91}Co_{0.09}O$	463	3.075	11.621	6.061	1.781	1.708	1.917	5.219	40	15.7
$Zn_{0.86}Co_{0.14}O$	722	2.993	11.872	6.192	1.778	1.708	1.917	5.000	96	16.0
$Zn_{0.82}Co_{0.18}O$	396	3.023	19.290	5.947	2.166	2.060	3.243	9.171	74	/
$Zn_{0.78}Co_{0.22}O$	395	3.009	19.706	6.037	2.169	2.065	3.264	8.956	78	/

Table 3.
Dispersion parameters of the films extracted by fitting the experimental data with Eq.(7).

$$E_0^2 = \frac{M_{-1}}{M_{-3}} \tag{9}$$

$$E_d^2 = \frac{M_{-1}^3}{M_{-3}} \tag{10}$$

These moments represent the measure of the average bond strength. The two moments M_{-1} and M_{-3} were calculated from the data of E_0 and E_d and are given in **Table 3**.

The calculated refractive indices of ZnO and $Zn_{1-x}Co_xO$ films (**Figure 4**) exhibit a function of the wavelength. It is found that the refractive indices at 598 nm of ZnO, $Zn_{0.95}Co_{0.05}O$ and $Zn_{0.78}Co_{0.22}O$ films are equal to 1.77, 1.96 and 2.16, respectively.

It can be noticed that the above calculated refractive indices are equal or a little greater than that of ZnO film prepared under the same conditions. This might be due to the fact that the index of refraction is sensitive to structural defects (for example voids, dopants, inclusions), thus it can provide an important information concerning the microstructure of the material.

It can be observed that the refractive indices measured above are equal to, or slightly greater than, the ZnO film prepared under the same conditions. This may be due to the fact that the refractive index is vulnerable to structural defects (such as voids, dopants, inclusions), so it can provide valuable details about the material's microstructure. As Zn(CH3COO)2 was oxidized into ZnO, gases like CH3COOH, H2O, etc. were manufactured. Consequently, because of the release of these gases, pores can be easily formed. Using the LorentzLorenz Equation [32], porosity P is determined from optical constants:

$$P = 1 - \frac{\left(n_{film}^2 - 1\right)\left(n_{bulk}^2 + 2\right)}{\left(n_{film}^2 + 2\right)\left(n_{bulk}^2 - 1\right)} \tag{11}$$

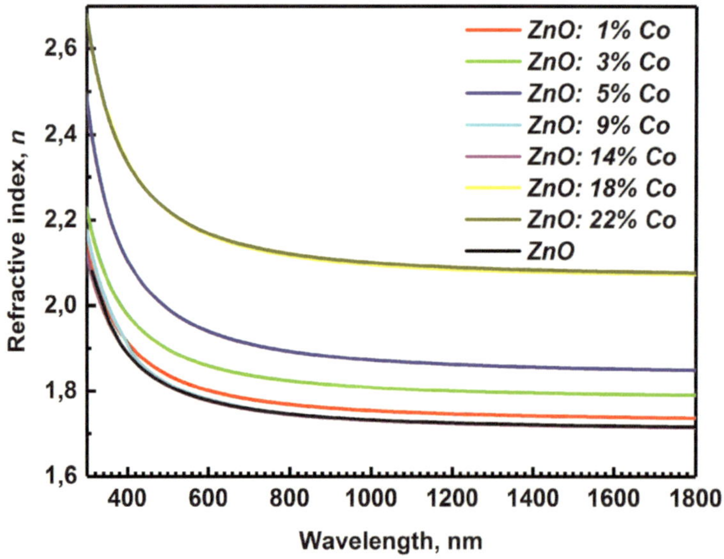

Figure 4.
Refractive index of $Zn_{1-x}Co_xO$ films grown on glass substrate at T_S = 450 °C.

In which the n_{film} value (1,771 at 598 nm) represents the refractive indices of porous ZnO films, the nbulk represents the ZnO bulk refractive indices at the same wavelength of 1,996. The average mass density of the film ρ_{film} is related to the porosity (P) and bulk density (ρ_{bulk}) of ZnO through Eq. (12):

$$\rho_{film} = \rho_{bulk}(1 - P) \tag{12}$$

Against the bulk density ρ_{bulk} = 5.61 g.cm^{-3}, we calculated P = 0.1659 and ρ_{film} = 4.68 g.cm^{-3}. The concentration of cobalt cations N_{Co} for x at. doping level in the films can be measured as:

$$N_{Co} = \frac{\rho_{film} N_{Av}}{M} \times x \tag{13}$$

where N_{Av} is the Avogadro constant and M the molar mass. With the values ρ_{film} = 4.68 g cm^{-3}, N_{Av} = 6.022 × 10^{23} mol^{-1}, and molar mass for ZnO, M = 81.408 g.mol^{-1}, an example (x = 0.05) of the calculated value (x = 0.05) of N_{Co} is 1.73 × 10^{21} cm^{-3}.

As a tool for calculating the concentration of impurities in a host from known values and calculated absorption coefficients, oscillator intensity is also used. Classically, the power of the oscillator f is the number of electric dipole oscillators that can be simulated (inthe dielectric dipole approximation) by the radiation field and has a value close to one for strongly permitted transitions. Integrated optical transition absorption is connected by the well known Smakula formula [33] to the concentration of absorbing centers N, refraction index n, and oscillator intensity f:

$$Nf = 8.21 \times 10^{16} cm^{-3} \frac{n}{(n^2 + 2)^2} \int \alpha(E) dE \tag{14}$$

where n is the refractive index of intersubband transitions, α is the decadic absorption coefficient in cm^{-1} and E is the energy in eV. For Gaussian absorption bands the integral is:

$$\frac{1}{2}\sqrt{\frac{\pi}{\ln 2}} \alpha_{\max} W \tag{15}$$

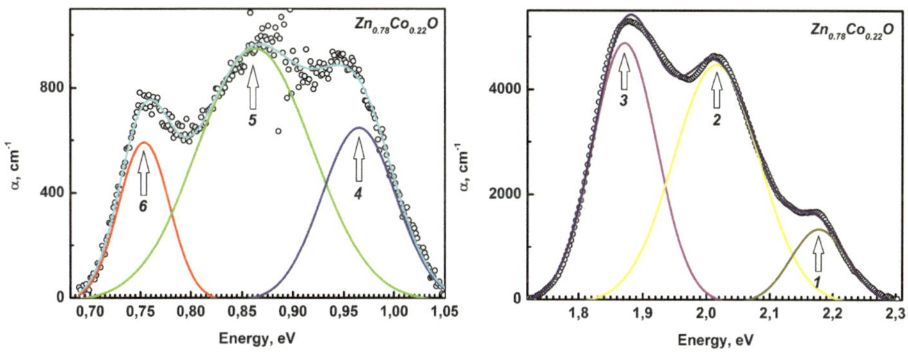

Figure 5.
The Gaussian deconvolution of the absorption coefficient of $Zn_{0.78}Co_{0.22}O$ films deposited on glass substrate at 450 °C.

N° Peak	E_c (eV)	w (eV)	A (eV/cm)	sigma	FWHM (eV)	height (cm^{-1})
1	$2.17774 \pm 6.04092 \times 10^{4}$	0.08866 ± 0.00118	156.38794 ± 3.46173	0.04433	0.10439	1407.38652
2	$2.01508 \pm 3.94746 \times 10^{-4}$	0.13482 ± 0.00125	765.61193 ± 6.88833	0.06741	0.15874	4530.94569
3	$1.87161 \pm 3.34026 \times 10^{-4}$	$0.10097 \pm 4.8835 \times 10^{-4}$	625.04657 ± 6.30209	0.05048	0.11888	4939.46937
4	0.96495 ± 0.00109	0.07376 ± 0.00166	60.94398 ± 3.91289	0.03688	0.08684	659.27169
5	0.86261 ± 0.00123	0.11265 ± 0.00363	135.47616 ± 4.41433	0.05633	0.13264	959.54823
6	$0.75363 \pm 3.62511 \times 10^{-4}$	0.05038 ± 0.00103	38.10798 ± 1.62028	0.02519	0.05932	603.52364

Table 4.
Deconvolution results of the absorption coefficient of the $Zn_{0.78}Co_{0.22}O$ $\alpha(cm^{-1}) \infty \frac{A}{w\sqrt{\pi/2}} \exp\left(-2\left(\frac{E-E_c}{w}\right)^2\right)$ *sample by the function.*

with maximum absorption α and full width at half maximum *W*. Eq. (14) can be expressed as follow:

$$Nf = 8.74 \times 10^{16} cm^{-3} \frac{n}{(n^2+2)^2} \alpha_{\max} W \tag{16}$$

Due to the Co^{2+} d-d transitions, it is difficult to measure the absorbance because the total transmittance value is different for each film. The absorption coefficient ($\alpha \approx (1/d) \times \ln(1/T)$) was then used because it is normalized by the thickness of the

	1	2	3	4	5	6
$Zn_{0.99}Co_{0.01}O$: d = 1350 nm, N_{Co}^{cal} = 3.403 × 10^{20} cm^{-3}, N_{Co}^{exp} = 3.622 × 10^{21} cm^{-3}, $\Sigma f_{i\to j}$ = 6.060 × 10^{-3}						
Refractive index	1.811	1.799	1.789	1.745	1.742	1.739
$f_{i\to j}$, ×10^{-3}	0.161	1.472	1.452	1.481	1.481	0.010
N, ×10^{20} cm^{-3}	11.16	1.133	3.069	1.806	0.746	18.310
$Zn_{0.97}Co_{0.03}O$: d = 846 nm, N_{Co}^{cal} = 9.835 × 10^{20} cm^{-3}, N_{Co}^{exp} = 1.223 × 10^{21} cm^{-3}, $\Sigma f_{i\to j}$ = 0.025						
Refractive index	1.869	1.856	1.845	1.798	1.795	1.792
$f_{i\to j}$, ×10^{-3}	0.597	5.803	4.696	6.934	5.998	1.039
N, ×10^{20} cm^{-3}	6.514	1.533	2.506	1.431	0.046	0.205
$Zn_{0.95}Co_{0.05}O$: d = 421 nm, N_{Co}^{cal} = 1.867 × 10^{21} cm^{-3}, N_{Co}^{exp} = 1.393 × 10^{20} cm^{-3}, $\Sigma f_{i\to j}$ = 0.100						
Refractive index	1.953	1.935	1.921	1.858	1.855	1.851
$f_{i\to j}$, ×10^{-3}	9.665	19.88	19.48	22.11	22.31	7.224
N, ×10^{19} cm^{-3}	2.531	5.545	4.429	0.209	0.383	0.835
$Zn_{0.91}Co_{0.09}O$: d = 463 nm, N_{Co}^{cal} = 2.876 × 10^{21} cm^{-3}, N_{Co}^{exp} = 4.887 × 10^{20} cm^{-3}, $\Sigma f_{i\to j}$ = 0.178						
Refractive index	1.791	1.778	1.768	1.722	1.720	1.717
$f_{i\to j}$, ×10^{-3}	15.10	38.86	38.87	38.45	38.87	8.333
N, ×10^{20} cm^{-3}	1.533	1.530	1.290	0.214	0.065	0.255
$Zn_{0.86}Co_{0.14}O$: d = 722 nm, N_{Co}^{cal} = 4.706 × 10^{21} cm^{-3}, N_{Co}^{exp} = 6.307 × 10^{20} cm^{-3}, $\Sigma f_{i\to j}$ = 0.174						
Refractive index	1.786	1.775	1.765	1.722	1.720	1.717
$f_{i\to j}$, ×10^{-3}	6.730	29.53	31.38	36.31	38.78	31.206
N, ×10^{20} cm^{-3}	2.532	1.857	1.389	0.108	0.305	0.116
$Zn_{0.82}Co_{0.18}O$: d = 396 nm, N_{Co}^{cal} = 6.392 × 10^{21} cm^{-3}, N_{Co}^{exp} = 3.327 × 10^{20} cm^{-3}, $\Sigma f_{i\to j}$ = 0.268						
Refractive index	2.180	2.161	2.146	2.081	2.078	2.075
$f_{i\to j}$, ×10^{-3}	17.92	29.71	41.94	74.64	90.80	13.161
N, ×10^{20} cm^{-3}	0.931	1.173	0.845	0.205	0.027	0.146
$Zn_{0.78}Co_{0.22}O$: d = 395 nm, N_{Co}^{cal} = 7.263 × 10^{21} cm^{-3}, N_{Co}^{exp} = 1.791 × 10^{20} cm^{-3}, $\Sigma f_{i\to j}$ = 0.417						
Refractive index	2.182	2.164	2.149	2.085	2.082	2.078
$f_{i\to j}$, ×10^{-3}	22.72	36.66	41.92	106.3	111.3	98.643
N, ×10^{20} cm^{-3}	0.269	0.830	0.600	0.024	0.052	0.016

1: 567 nm ($^4A_2(F) \to {}^2E(G)$); 2: 610 nm ($^4A_2(F) \to {}^4T_1(P)$); 3: 657 nm ($^4A_2(F) \to {}^2A_1(G)$); 4: 1297 nm ($^4A_2(F) \to {}^4T_1(F)$); 5: 1410 nm ($^4A_2(F) \to {}^4T_1(F)$); 6: 1647 nm ($^4A_2(F) \to {}^4T_1(F)$).

Table 5.
Level of absorbing sites N and oscillator strength f of Co^{2+} ions transition d-d.

film (d) as shown in **Figure 5**. The mathematical treatment of the coefficient of absorption is seen. It was possible to achieve a good fit for the multi-peak combination by optimizing the peak position and half-width of the Gaussian peaks.

At the bottom of **Figure 5**, the Gaussian peaks (dashed lines) while the solid line reflects the linear combination with a constant history of the multi-Gaussian peaks. **Table 4** shows the Gaussian peak position, field, width (eV) and height (α_{max}, cm^{-1}). The mathematical treatment of the absorption coefficient has shown that a series of overlapping bands consists of a large visible and near infrared spectral area. 0.75, ~0.86, ~0.96, ~1.87, ~22 are distinguished by six dominating bands.

Understanding the strengths of the oscillator $f_{i \to j}$ as determined from Eq. (7) The refractive index value of the intersubband transitions, i.e. at $\lambda_{i \to j}$ film, α_{max} and full width at half maximum W as defined by the Gaussian deconvolution of the absorption coefficient, enables the concentration of absorbing centers N. to be calculated from the formula of Smakula.

Table 5 displays the values obtained for the concentration of absorbing centers (N) and oscillator power (f) of the fingerprint of d-d transitions of Co^{2+} ions at the T_d symmetry sites.

For the investigated films, the amount of the oscillator intensity ($\Sigma f_{i \to j}$) from ground state $^4A_2(F)$ to all other states ranges from 0.006 to 0.417.

As mentioned above, the values of the direct optical band-gap is reduced from 3.26 to 3.00 eV. This significant band-gap reduction is due to enhanced Co^{2+} ions incorporation in the films as confirmed by the obtained concentration of absorbing centres.

In the deposition method by ultrasonic spray pyrolysis, the film growth is carried out by thermal decomposition of a precipitate at the substrate; This deposit results from the vaporization of the aerosol droplets. In this situation, the material that forms contains different types of impurities causing a disorder in the structure. In this case, the crystal lattice bounded by E_V and E_C may disappear. When the disorder becomes too high (eg with the appearance of dangling bonds or impurities in the material), the tails may encroach. We define the notion of Urbach parameter (E_{Urb}) to characterize this disorder. It is possible to estimate the existing disorder in

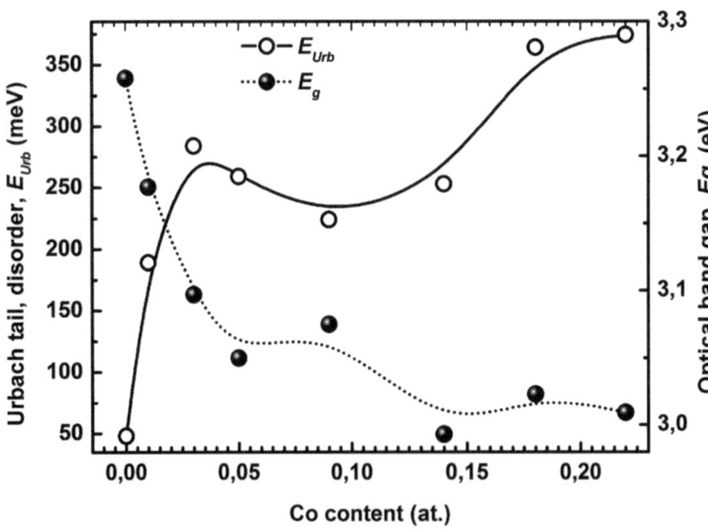

Figure 6.
Doping result and the variation of the bandgap and Urbach energy for different Co content.

the layers by studying the variations of the absorption coefficient. Indeed, the absorption coefficient can be expressed by the Equation [34]:

$$\alpha = \alpha_0 \exp\left(\frac{h\upsilon}{E_{Urb}}\right), \qquad h\upsilon < E_g$$

All our experimental results for the variation of the optical bandgap of the thin layers and disorders (Urbach energy of each sample is shown) as function of cobalt contents is shown in **Figure 6**. It is observed that the bandgap decreases with increasing cobalt content. The presence of high concentrations of localized states in the thin films is responsible for the reduction in the width of optical bandgap. Therefore, the addition of Co increases the concentration of localized states in the thin film leading to the decrease of the bandgap.

4. Conclusion

X-ray diffraction analysis using the Rietveld method shows that the as-deposited ZnO and $Zn_{1-x}Co_xO$ (x = 0.01–0.22) films are pure single wurtzite phase. The lattice parameter "*a*" increases with Co content while "*c*" seems to not much affected, thereby the volume of unit cell increases with increasing Co. However, the evolution of lattice parameters with Co content, is more complex than expected.

An optical model, which combines the Wemple-DiDomenico model, absorption coefficient of an electronic transition and Tauc-Urbach model, has been proposed to simulate the optical constants and thicknesses of Co doped ZnO films from normal incidence transmittance. The simulated transmission is found to be well matched to the measured transmission. The dispersion curves of the refractive index follow the model of the single oscillator. The dispersion parameters and the optical constants of the layers have been determined. The concentration of absorbing centres N_{Co} and oscillator strength f of d–d transition of Co^{2+} ions are also calculated from Smakula's formula.

The presence of high concentrations of localized states in the thin films is responsible for the reduction in the width of optical bandgap.

Acknowledgements

The authors would like to thank the National Project Research (PNR) and LASPI^{2}A Laboratory of Khenchela University (Algeria) for their financial support of this research project.

Author details

Sabrina Roguai[1*] and Abdelkader Djelloul[2]

1 LASPI^{2}A Laboratory of Structures, Properties and Inter-Atomic Interactions, Abbes Laghrour University, Khenchela, Algeria

2 Science of Matter, Abbes Laghrour University, Khenchela, Algeria

*Address all correspondence to: rog.sabrina@yahoo.fr

References

[1] Oktik S, Cost Low. non-vacuum tech niques for the preparation of thin/thick films for photovoltaic applications.Prog. Growth Charact. 1988; 17:171–240. doi : org/10.1016/0146-3535(88)90006-8

[2] Bouzid K., Djelloul A., Bouzid N, Bougdira J, Electrical resistivity and photoluminescence of zinc oxide films prepared by ultrasonic spray pyrolysis. Phys. Status Solid. A. 2009; 206:106–115. doi.org/10.1002/pssa .200824403

[3] Janotti A, Van de Walle Ch-G, Fundamentals of zinc oxide as asemiconductor. Rep. Prog. Phys. 2009; 72: 126501.

[4] Ellmer K, Klein A, Rech-(eds.) B, Transparent Conductive Zinc Oxide-Basics and Applications in Thin Film Solar Cells. Series: Springer Series in Materials Science.2008;104.

[5] Yang HM, Nie S. Preparation and characterization of Co doped ZnO nanomaterials. Mater Chem Phys. 2009; 114: 279–282. doi: org/10.1016/j. matchemphys.2008.09.017

[6] Yang M., Guo ZX., Qiu KH, et al. Synthesis and characterization of Mn-doped ZnO column arrays. Appl Surf Sci. 2010;256: 4201–4205. doi :org/ 10.1016/j.apsusc.2010.01.125

[7] Saal H, Bredow T, Binnewies M. Band gap engineering of ZnO via doping with Manganese effect of Mn clustering. Phys Chem Chem Phys. 2009; 11: 3201–3209. doi :org/10.1039/B901596E

[8] Kumar GM, Ilanchezhiyan P, Kawakita J, et al. Magnetic and optical property studies on controlled low-temperature fabricated one dimensional Cr doped ZnO nanorods. Cryst Eng Commun. 2010; 12 :1887–1892. doi :org/ 10.1039/B924643F

[9] Fabbiyola S, Kennedy L-J, Aruldoss U, Bououdina M, Dakhel A-A, Judith Vijaya J, Synthesis of Co-doped ZnO nanoparticles via co-precipitation: structural, optical and magnetic properties. Powder. Technol. 2015; 286: 757–765. doi.org/10.1016/j.powtec. 2015.08.054

[10] Dietl T, Ohno H, Matsukura F, et al. Zener Model Description of Ferromagnetism in Zinc-Blende Magnetic Semiconductors. Science 2000 ;287 :1019–1022. doi: 10.1126/ science.287.5455.1019

[11] Wang YX, Ding X, Cheng Y, et al. Properties of Co-doped ZnO films prepared by electrochemical deposition. Cryst Res Technol. 2009;44(5):517–520. doi :org/10.1002/crat.200800466

[12] Song C, Zeng F, Geng KW, et al. The magnetic properties of Co-doped ZnO diluted magnetic insulator films prepared by direct current reactive magnetron co-sputtering. J Magn Magn Mater. 2007;309: 25–30.doi.org/ 10.1016/j. jmmm.2006.06.012

[13] Zukova A, Teiserskis A, Van Dijken S, et al. Giant moment and magnetic anisotropy in Co-doped ZnO films grown by pulse-injection metal organic chemical vapor deposition. Appl. Phys. Lett. 2006; 89:232503–232505. doi :org/10.1063/1.2399939

[14] Matsui H, Tabata H. Simultaneous control of growth mode and ferromagnetic ordering in Co-doped ZnO layers with Zn polarity. Phys Rev B. 2007; 75:014438–014447. doi : org/10.1103/PhysRevB.75.014438

[15] Sivagamasundari A, Pugaze R, Chandrasekar S, Rajagopan S, Kannan R, Absence of free carrier and paramagnetism in cobalt-doped ZnO nanoparticles synthesized at low

temperature using citrate sol-gel route. Appl. Nanosci. 2013;3: 383–388. DOI 10.1007/s13204-012-0146-0.

[16] Iqbal G, Faisal S, Khan S, Shams D-F, Nadhman A, Photo-inactivation and efflux pump inhibition of methicillin resistant *Staphylococcus aureus* using thiolated cobalt doped ZnO nanoparticles. Journal of Photochemistry & Photobiology. B: Biology.2019;192: 141–146. doi.org/10.1016/j.jphotob iol.2019.01.021

[17] Sindhu H-S, Kulkarni D-S, Choudhary R-J, Babu B-D, Rajendra B-V, Influence of cobalt doping on structure, optical and magnetic properties of spray pyrolysed nano structured ZnO films, Physica B: Physics of Condensed Matter.2019;578:18–26 d oi.org/10.1016/j.physb.2019.07.034

[18] Ivill M, Pearton S-J, Rawal S, Leu L, Sadik P, Das R, Hebard A-F, Chisholm M, Budai John-D, and Norton David-P , Structure and magnetism of cobalt doped ZnO thin films. New Journal of Physics. 2008;10:065002

[19] Kohan A-F, Ceder G, Morgan D, Van de Walle C-G, First-principles of native point defects in ZnO. Physical Review B. 2000; 61: 15019. doi.org/ 10.1103/PhysRevB.61.15019

[20] Roguai S, Djelloul A., Nouveau C, et al. Structure, microstructure and determination of optical constants from transmittance data of co-doped $Zn_{0.90}Co_{0.05}M_{0.05}O$ (M = Al, Cu, Cd, Na) films. J. Alloys Compd.2014; 599:150–158 doi : org/10.1016/j.jallc om.2014.02.080

[21] Bouloudenine M, Viart N, Colis S, Kortus J, Dinia A. Antiferromagnetism in bulk $Zn_{1-x}Co_xO$ $Zn_{1-x}Co_xO$ magnetic semiconductors prepared by the coprécipitation technique. Appl. Phys. Lett.2005; 87: 052501. doi.org/10.1063/ 1.2001739

[22] Chen X-C, Zhou J-P, Wang H-Y, Xu P-S, Pan G-Q, Chin. Phys. B.2011; 20:9.

[23] Bao D, Gu H, Kuang A, Sol-gel-derived c-axis oriented ZnO thin films. Thin Solid Films. 1998;312,:37–39

[24] Benramache S, Benhaoua B, Influence of substrate temperature and Cobalt concentration on structural and optical properties of ZnO thin films prepared by Ultrasonic spray technique. Superlattices and Microstructures. 2012; 52: 807–815 doi.org/10.1016/j. spmi.2012.06.005.

[25] Wemple SH, DiDomenico M. Behavior of the Electronic Dielectric Constant in Covalent and Ionic Materials. Phys. Rev. B. 1971; 3: 1338–1351. DOI:https://doi.org/10.1103/Ph ysRevB.3.1338

[26] Tauc J, in: F. Abelès (Ed.), Optical Properties of Solids, North-Holland, Amsterdam, London.1972: 277–313.

[27] Davis EA, Mott NF. Conduction in non-crystalline systems V. Conductivity, optical absorption and photoconductivity in amorphous semiconductors. Philosophical Magazine. 1970; 22: 903–922. https://d oi.org/10.1080/14786437008221061

[28] http://refractiveindex.info/?group= GLASSES&material=BK7.

[29] Ashour A, Kaid MA, El-Sayed NZ, Ibrahim AA. Physical properties of ZnO thin films deposited by spray pyrolysis technique. Appl. Surf. Sci.2006; 252: 7844–7848. https://doi.org/10.1016/j.a psusc.2005.09.048

[30] Lommens P, Smet PF, Donega CM, et al. Photoluminescence properties of Co^{2+}-doped ZnO nanocrystals.J. Lumin. 2006 ;118 : 245–250. doi : org/10.1016/j. jlumin.2005.08.020

[31] Pereira AS, Ankiewicz AO, Gehlhoff W, et al. Surface modification

of Co-doped ZnO nanocrystals and its effects on the magnetic properties. J Appl Phys. 2008 ;103: 07D140. doi: 10.1063/1.2833300

[32] Baklanov MR, Mogilnikov KP, Polovinkin VG, Dultsev FN. Determination of pore size distribution in thin films by ellipsometric porosimetry. J. Vac. Sci. Technol. B. 2000;18: 1385–1391. **https://doi.org/10.1116/1.591390**

[33] Smakula A. Über Erregung und Entfärbung lichtelektrisch leitender Alkalihalogenide. Zeitschrift für Physi*k*. 1930; 59: 603–614.

[34] Pankove JI. Absorption Edge of Impure Gallium Arsenide. Phys. Rev. 1965;140: A 2059. DOI:https://doi.org/10.1103/PhysRev.140.A2059

Chapter 6

Multifunctional ZnO Nanoparticle: Based Coatings for Cultural Heritage Preventive Conservation

Ludmila Otilia Cinteză and Maria Antonia Tănase

Abstract

Coatings based on nanoparticles embedded in various filmogenic materials are still a hot topic in nanomaterial research, due to the exceptional variety of applications. The chapter present recent progress in synthesis and characterization of hybrid material with ZnO nanoparticles and their use as functional coatings for various substrates. The antibacterial and UV protection efficiency of ZnO nanoparticle-based coatings on paper and stone are discussed, with particular emphasize on the specific requirements for application in cultural heritage preventing conservation. Functional materials based on ZnO are presented as possible treatment for protection of historic and **archaeological** textiles and metal artifacts. Trends in environmental friendly methods to fabricate the ZnO nanoparticles will be evaluated, compared to classic ones, in terms of material characteristics and efficiency.

Keywords: zinc oxide nanoparticles, multifunctional coatings, antimicrobial, superhydrophobic, cultural heritage

1. Introduction

Defined as "design, production and application of structures, devices and systems by controlling shape and size at nanometer scale", nanotechnology is the most emerging field designed to revolutionize material science. Due to their unique properties, nanomaterials become an important resource for the scientific community in solving many issues related to many domains, from medicine to aeronautics, optoelectronics, clean energy or waste treatment. Among the multitude of metals and metal oxides that can be used in the fabrication of nanoparticles, zinc oxide (ZnO) is of a great importance due to its specific physical and chemical properties, easiness in processing, low cost and abundance of precursors. Zinc oxide is a semiconductor type material, with the band gap of 3.36 eV and an exciton-binding energy of 60 meV, making it a suitable nanomaterial for the optoelectronic industry [1]. At nanometric size, ZnO exhibits antibacterial, antifungal, UV-blocking or photocatalytic properties [2, 3]. Due to these properties it is well suited for applications in various fields such as biomedicine, textile industry, pharmaceutical and cosmetic industry, sensors, electronics, rubber industry [4, 5]. The antimicrobial properties of ZnO nanoparticles gain interest in the last decades and recommend them as replacement to classic biocidal treatments for the surfaces. Their antifungal and antibiofilm activities were investigated in the last two decades as potential tools in preventive conservation of cultural heritage monuments [6].

Due to their unique physical and chemical properties Zinc oxide nanoparticles have become a key material in developing new nanotechnologies suitable for coating of different substrates. As seen from the plethora of synthesis routes and materials used as reagents, this kind of nanomaterials have a large variety of morphologies and sizes which lead to unique properties. Multifunctional ZnO nanomaterials could represent an interesting choice to be used in protecting the cultural heritage, besides their industrial applications, demonstrating their great versatility and offering multiple benefits in this field.

2. Synthesis of zinc oxide nanoparticles

Extensive use of ZnO nanoparticles (NPs) generate important attention to their synthesis route, in order to achieve better control of size and shape of the obtained material. There are many methods that can be used such as: sol–gel, chemical vapor deposition, thermal oxidation [7], etc. but the easiest and most frequently used are the hydrothermal and solvothermal ones. Briefly, the precursor (zinc salt solution and a base solution) are mixed at different molar ratios in suitable solvents (water or nonpolar organics). The as obtained precipitate is washed and dried. Sometimes polymers are added to stabilize the obtained nanoparticles. By varying the solvent, precursors, molar ration or other reaction parameters one can obtain different size and morphologies of the nanoparticles.

Among the various synthesis methods, the hydrothermal one brings the most advantages such as low temperatures, low aggregation levels or good control over particle morphology and size. Synthesis of ZnO NPs with different morphologies using a simple hydrothermal method was reported [8]. Flowerlike nanoparticles are synthesized using Zinc nitrate ($Zn(NO_3)_2{\cdot}6H_2O$) and cetyltrimethyl ammonium bromide (CTABr,) dissolved in deionized water under magnetic stirring at room temperature. NaOH is then slowly added drop wise. The resultant solution is then added in a stainless-steel Teflon autoclave. The white precipitate obtained is centrifuged, and dried, obtaining flower-like structures. Same synthesis route in the absence of CTABr and NaOH is used to obtain nanoplates and nanoparticles. The XRD and SEM characterization showed wurtzite hexagonal structures, with the average diameter in the range of 40–80 nm for the nanoparticles, 30 nm thick nanoplates, and, nanoplates agglomeration forming flower-like structures of 2 μm diameter.

Complex three-dimensional nanoaggregates (**Figure 1**) with different flower-like morphologies, such as cauliflower-like, safflower-like microspheres and ixora-like nano-structures can be obtained by the hydrothermal method, at different concentrations of OH^- ions and Zn^{2+}-glutamic acid molar ratio [9].

Figure 1.
SEM images of the ZnO 3D nanoaggregates (a, (cauliflower); b, (safflower); c, (ixora); [9].

By varying the base added to the precursor solution, Ekthammathat et al. [10] obtain rod-like, pencil-like and star-like particles. Zinc foils are used as solid reagent and substrates for the direct growth of ZnO films. NaOH, LiOH and NH_4OH solutions in deionized water are used as source of OH^- ions. The hydroxide are introduced in Teflon stainless-steel autoclaves, and the clean Zn foils were added to each solution. The as-synthesized films on Zn substrates were rinsed and dried. When using NaOH as precursor, 100 nm diameter and 500 nm length nanorods were obtained, pencil like nanostructures of 300 nm diameter and more than 800 nm length in the case when using LiOH and in the case of NH_4OH the authors obtained star-like nanostructures of 200 nm diameter and 1 μm length.

Stanković et al. [11] studied the influence of the stabilizing agents on the shape of ZnO NPs obtained from zinc acetate, $Zn(CH_3COO)_2{\cdot}2H_2O$ and NaOH as precursors. In the case when polyvinyl pyrrolidone (PVP) is used as stabilizing agent, a mixture of hexagonal prismatic rods and incompletely formed hexagonal pyramid-like structures is obtained, with length approximately 1 μm and diameter of about 100 nm. Different morphologies were obtained when using polyvinyl alcohol (PVA). The particles are spherical and well dispersed with average particle diameter around 30 nm, while ellipsoid shaped structures with the length of 500–600 nm and diameter of about 100 nm were obtained when using poly α,γ, l-glutamic acid (PGA). The morphology of ZnO nanopowders obtained in hydrothermal synthesis is also influenced by the variation of pH value. Kumaresan et al. [12] synthesized ZnO microparticles from $ZnCl_2$ and NH_4OH at different pH (7, 9, 11 and 13). When the pH is fixed at 7, the XRD and Field Emission Scanning Electron Microscope (FESEM) analysis reveal hexagonal shaped ZnO nanorods arranged in a flower like structure. The length and width of the nanorods which form nanorod-flower structure is about 4790 nm and 1000 nm respectively. When increasing the pH to 9, spheroidal and hexagonal disk shaped morphologies were obtained. Nanorods of various lengths and widths were observed when the pH value rise to 11, whereas in the case of pH = 13, nanoflower structures of 4.8 μm diameter can be observed.

The temperature, pressure and reaction time can be minimized using the microwave assisted hydrothermal synthesis [13]. Zinc oxide nanorods were produced from $ZnCl_2$ and hydrazine as reagent. The white precipitate initially formed is transferred in a Teflon container and the mixture was irradiated and then cooled at room temperature, centrifuged, rinsed and dried. The nanorods formed are polydispersed, with diameter between 65.14 and 170.93 and the length between 241.49 and 941.16 nm. Hasanpoor [14] reports both the effect of the precursors and of the radiation power over the size and shape of ZnO nanostructures. The ZnO powder prepared from the solution of $Zn(NO_3)_2{\cdot}6H_2O$ and ammonia to reach the pH 11.5, irradiated for 15 min at 510 W exhibit needle-shaped nanostructures. When the mixture was subjected to irradiation for 10 min at 680 W, the particles were flower like.

Zare et al. [15] proposed a solvothermal synthesis in the presence of surfactants to obtain ZnO nanoparticles of different morphologies. Methanol solutions of $Zn(NO_3)_2{\cdot}6H_2O$ and NaOH were mixed under constant stirring. Then surfactant solutions were added and the new mixture was transferred to a stainless-steel autoclave which was further heated and cooled at room temperature. The final product was washed and dried. When the surfactant was oleic acid, the obtained product has flower-like morphology (102 nm), when using gluconic acid the product was a snowflake structure (119 nm) and when using Tween 80 the morphology was granular, with quasi spherical nanoparticles (69 nm). A solvothermal and hydrothermal microwave assisted synthesis of ZnO nanoparticles are compared by Krishnapriya [16]. Two zinc precursors, $[Zn(NH_3)_4]^{2+}$and $[Zn(OH)_4]^{2-}$ were mixed with double distilled water, ethylene glycol and ethanol. The as obtained mixtures are added to a quartz vessel and heated in the microwave, thancooled to room temperature and the white precipitate obtained was filtered, washed, centrifuged and dried. The obtained nanoparticles have different flower-like morphologies, depending on the polarity of the solvent and on the precursor concentration. The diameter of the obtained nanoparticles is between 10 and 500 nm and their length between 300 nm and 3 μm.

Solvothermal methods can also be made in supercritical fluids. In a study done by our group [17], flowerlike ZnO nanoparticles were synthesized using an adapted method in supercritical CO_2. The reagent Zn $(NO_3)_2$, NaOH and cetyltrimethylammonium ammonium bromide (CTABr) were dissolved in a mixture of ethanol and water. The reaction is performed in high temperature and pressure conditions, in a stainless-steel autoclave with Teflon coating. In the modified synthesis in supercritical CO_2, the reaction cell was filled with carbon dioxide at certain temperature and pressure and remained at these conditions for the various period of time. After the system was cooled at −10°C was slowly depressurized. The white precipitates obtained in various conditions were collected, filtered and washed several time in water and alcohol. Both products obtained using the solvothermal synthesis performed in the autoclave and in the case of supercritical CO_2 variation were flower like aggregates of 700–900 nm, as seen in the SEM images presented in **Figure 2**.

In the last decade attention has been focused on using the green synthesis approach for nanoparticles in order to reduce the drawbacks of the classic methods

Figure 2.
SEM images of the ZnO nanopowders: (A) solvothermal synthesis in normal conditions, without surfactant; (B) solvothermal synthesis in autoclave, in the presence of surfactant CTABr; (C) supercritical CO2 assisted solvothermal synthesis, in the presence of surfactant CTABr (lowmagnification); (D) high magnification of sample C [17].

No.	Vegetal source	Precursor	Size(nm)	Morphology	Reference
1	Laurul Nobilis L.	Zn Nitrate Zn Acetate	25.26 21.49	Flowerlike Bullet like	[19]
2	Pineapple	Zn Nitrate	73–123	Flower rod	[20]
3	*Garcinia mangostana*	Zn Nitrate	21	Spherical and hexagonal	[21]
4	Deverra tortuosa	Zn Nitrate	9.26–31.18	Spherical	[22]
5	*Calotropis procera*	Zn Nitrate	15–25	Spherical	[23]
6	Protoparmeliopsis muralis	Zn Nitrate	178.06	Spherical	[24]
7	Costus pictus D. Don	Zn Nitrate	20–80	Hexagonal rod	[25]
8	*Prunus dulcis*	Zn Nitrate	25.2	Spherical	[26]
9	*Citrus sinensis*	Zn Nitrate	10–20	Spherical	[27]
10	*Phoenix dactylifera*	Zn Nitrate	31.6	Spherical	[28]
11	*Matricaria chamomilla* L	Zn Oxide	49.8–191	Hexagonal rods	[29]
12	*Citrus Limon*	Zn Oxide	60.8	Hexagonal prism	[30]
13	*Mentha spicata*	Zn Acetate	74.68	Spherical	[31]
14	Oak Tree	Zn Acetate	34	Spherical	[32]
15	*Tecoma castanifolia*	Zn Sulfate	70–75	Spherical	[33]

Table 1.
Green synthesis of ZnO nanoparticles.

discussed above (mainly the large use of toxic chemicals). The green approach implies the use of microorganisms (bacteria, fungus, algae) or plants in the synthesis of various nanoparticles [18]. Briefly, in this synthesis approach, a precursor molecule (zinc acetate more often) and the plant extract are mixed under stirring in order to form Zn^{2+} ions which are transformed into ZnO nanoparticles in a hydrolysis reaction. Some recent results on green synthesis of ZnO nanoparticles are summarized in **Table 1**.

The main advantage is that there is no need of any stabilizing agents, while phytochemicals (polysaccharides, polyphenolic compounds, vitamins, amino acids, etc) content from the plant act both as reducing agent and capping or stabilizing agent [18].

3. Properties of zinc oxide nanoparticles

As semiconducting materials, Zinc oxide nanoparticles were first used in the optoelectronic industry and for their catalytic properties. As it is shown in the previous sections, there are many ways of synthesizing ZnO nanopowders in a large variety of dimensions and morphologies. These two main characteristics dictate the properties and further uses of zinc oxide nanoparticles.

3.1 Antimicrobial activity

One of the most important properties of ZnO NPs is their ability to inhibit the growth of various bacterial and fungal strains, both in solution or on the surfaces. Although their mechanism of action is not fully understood, the researchers proposed two possible routes: the physical and chemical interactions. In terms of physical interactions, plasma membrane disruption, cellular internalization or mechanical damage are the most frequent mechanisms, while in the case of chemical interactions there are three possible ways: reactive oxygen species (ROS) production and Zn^{2+} release and the photoproduction of H_2O_2 [34].

Antifungal activity of ZnO NPs against Asperillus niger [35], a microorganism responsible for the deterioration of a large variety of common or historic objects was investigated. The material with antimicrobial effect was obtained by dissolving micrometric ZnO and urea in glycerol. The clusters obtained have a pseudo spherical morphology and possess hierarchical structures, with average size of 590 nm, while the pristine ZnO crystallites have 15–30 nm. The obtained product shows antifungal activity against fungus *A. niger* and also proved bactericidal activity against the bacteria *Escherichia coli* (*E. coli*) and *Staphylococcus aureus* (*S. aureus*). In both cases, the formation of bacterial colonies decreased at very low concetrations (3 ppm) of ZnO in the culture media.

Raghupathi et al. showed that the antibacterial activity of the ZnO nanopowders was inversely proportional to the size of the nanoparticles in the case of gram positive bacteria, *S. aureus* [36]. The tested ZnO NPs were spherical in shape, with size varying between 12 and 307 nm. To determine the antibacterial activity of the ZnO nanoparticles a wide range of microorganisms were studied: *S. aureus*, *S. epidermidis*, S. pyogenes, *E. faecalis*, *B. subtilis*, *B. cereus*, *E. coli*, S. typhimurium, P. alcaligenes, *E. aerogenes*, S. flexneri, etc. The patterns of growth inhibition of microorganisms were similar and 95% growth inhibition was achieved with a concentration of 6 mM colloidal suspension of zinc oxide nanoparticles.

Functional nanocoating [37] based on ZnO and carboxymethyl-chitosan, with antibacterial and UV protecting properties was reported. To obtain the bio nanocomposite a water-soluble solution of carboxymethyl-chitosan was prepared, and further used as reaction media to synthesized ZnO NPs, which were spherical with an average size of 100 nm. The antibacterial activity of the coating, applied in various concentrations on the cotton samples was measured against *S. aureus* and *E. coli*. The results show that all treated samples have inhibition zone larger than the untreated samples. As the concentration of the suspension increase, the zone of inhibition also increases. Also, the resistivity of *E. coli* is larger compared to *S. aureus*. The values of the UV Protection Factor (UPF) increased when increasing the temperature of curring of the nanocomposite treated fabrics. The same observations were made using UV transmittance tests. Unfortunately, the chitosan derivative was only used as stabilizing agent and polymer matrix, no study was performed to elucidate the role of biopolymer in the antibacterial activity of the nanocomposite.

The antimicrobial activity of ZnO NPs have been extensively investigated against many common bacterial and fungi, but systematic studies on the influence of size, shape, surface functionalization, specificity of strains are still missing. In particular, very few studies have been reported on the specific microorganism that colonize the historic monument or artistic artifacts.

3.2 Superhydrophobic functionalization of the surfaces

Another important property of zinc oxide nanoparticle is their ability to be used in superhydrophobic coating for different substrates. For a material to be considered superhydrophobic it must possess a certain roughness, with hierarchical structures in the micro and nano meter scale and the measured contact angle must be greater than 150°. Lee et al. [38] proposed a superhydrophobic nanocoating for various substrates, using a simple dispersion of ZnO nanoparticles in ethanol to ensure the suitable roughness. The colloidal suspension was spin-coated on a silicone substrate several times and immersed in an ethanolic solution of stearic acid. From SEM measurements, the as obtained ZnO nanopowders were spherical, with size ranging from less the 10 nm to hundreds of nm. The large distribution of particle size leads to a rough surface, the roughness increasing when the number of coating cycles increases. The wetting properties of the coated substrate were tested by measuring the water contact angle (WCA). The water CA of the untreated substrate was below 5°. After the chemical modification with the obtained nanocomposite the WCA increased beyond 150°, turning the material in superhydrophobic one. When the number of the coating cycles exceeded 20, the WCA dropped to 151°. All tested substrates (silicone, paper, cotton and polyethylene tereftalate) show similar behavior.

Zinc oxide based coatings with both superhydrophobic and antibacterial properties were synthesized by Shaban et al. [2]. The solvothermal method is used, with concentrations of zinc acetate dissolved in 2-methoxyethanol and mono ethanolamine as reagents. The pH of the solution was controlled using NaOH or acetic acid. Magnesium acetate tetrahydrate was added as dopant. Characterization methods showed that the morphology of the nanoparticles is cillindrical with the average diameter of 13.4 nm. The prepared ZnO doped dispersion was used for the treatment of cotton fabrics. The wettability of the coated textile was tested by water contact angle. Best results were obtained when the concentration of the zinc precursor was 0.5 M and the pH of the solution 7. The measured water contact angle was 154°, demonstrating the superhydrophobic character of the coating. The resistance of cotton samples against the attack of microorganisms *B. subtilis*, K. pneumonia, *E. coli*, and S. typhimurium was also investigated and all tested samples showed antibacterial properties.

The work of Ghasemi [39] present a facile one-step method for the preparation of a hybrid zinc oxide/octadecanethiol material with superhydrophobic and antibacterial properties that can be used as efficient coatings for cotton fibers. A fixed amount of octadecanethiol was mixed with zinc oxide NPs and chloroform. Fabric samples were immersed in this suspension for different amounts of time. Another approach was a two-step dip coating method: fabric samples were dipcoated in ZnO dispersion, removed and dried. The treated textile was then immersed in an octadecanethiol solution. As seen from SEM images one can observe some roughnesses on the surface of the fabrics, with the maximum cluster diameter of 2 μm. The wetting properties of the samples were determined by contact angle measurements. Pristine fabric and the samples coated with only ZnO or octadecanethiol showed hydrophilic behavior, while the textile samples coated with both ZnO dispersion and octadecanethiol exhibits superhydrophobic properties, with the measured WCA of 161°. Coated fabrics were tested for their antimicrobial properties by cultivating microorganisms *S. aureus* and *E. coli* on the surface of the textile materials. Results show that *S. aureus* severely colonized on the pristine fabric while the superhydrophobic sample was more resistant on the bacterial attack. The inhibition of the bacterial growth was more obvious in the case of *E. coli*, results corresponding to CFU measurements.

3.3 UV blocking and photocatalytic properties

Also it is worth mentioning the ability of zinc oxide to form UV absorbing and anticorrosion coatings. In a work developed by Becheri [40] ZnO NPs with UV-absorbing properties are synthesized using both hydrothermal and solvothermal method in order to obtain coating materials for cotton and wool fabrics. $ZnCl_2$ was dissolved in water or ethandiole and a solution of NaOH was added dropwise under stirring. The particles were separated from the supernantant, washed and peptized (to disrupt the micro aglomeration and release the nanounits of ZnO), washed again and calcinated. The as obtained nanoparticles had the average diameter of 21 nm (in the case of hydrothermal synthesis) and 9 nm (in the case of solvothermal synthesis). In both cases, the nanoparticles were spherical. Cotton and wool fabrics were conditioned and added to a 2-propanol dispersion of ZnO nanoparticles, dried and washed. The UV absorbing properties were tested by means of UV spectroscopy, UFP and UV transmittance. Untreated cotton does not absorb UV radiation, while the untreated wool absorbs in the 200–300 nm region. The application of ZnO nanoparticles on both cotton and wool increased the absorption of UV radiations over the entire investigated UV spectrum, with better results for the ZnO NPs obtained in the solvothermal synthesis, with smaller size.

ZnO based photocatalysts have been in the spotlight of research in the recent years because of their benefits such as low cost and high quantum efficiency. ZnO was found to decompose organic pollutants, for example dyes like Crystal Violet, Methylene Blue, Orange G or Methyl Orange at a faster rate compared to Degussa P25 (TiO_2 classic photocatalyst) under UV–visible light, indicating that the photosensitization of ZnO by dyes favored the visible light response with an enhanced charge carrier separation. Also, the discoloration of Methyl Orange was found to be greater for ZnO NP with smaller particle size [41].

3.4 Anticorrosion properties

ZnO was considered for many industrial applications due to its ability to act as anticorrosive protection for many metals. Superhydrophobic fluorinated polysiloxane-ZnO nanocomposite with corrosion resistance was developed [42] as multifunctional coating for metallic surfaces. The ZnO particles dissolved in butylacetate were vigorously stirred, followed by addition of stearic acid, washed and dried. Modified ZnO powder was added to an aqueous fluorinated α,ω-bis(hydrogen)-terminated poly (dimethylsiloxane) (FPDHS) dispersion and mixed in order to obtain the final nanocomposite (ZnO/FPDHS) which was deposited onto steel substrates. The water contact angle WCA of the nanocomposite was 166°. Polarization curves of the bare steel, FPDHS coating, modified ZnO coating and modified-ZnO/FPDHS coating were measured in % NaCl aqueous solution. It can be seen that the FPDHS coating shows better corrosion resistance than the bare steel, owing to its lower corrosion current density and corrosion rate. As expected, the modified-ZnO/FPDHS exhibits the lowest corrosion rate.

Despite the fact that ZnO NPs show spectacular properties, such as extended antibacterial activity on most species of microorganisms, ability to protect against corrosion, photocatalytic and UV blocking activities, they have been designed and tested in most cases for industrial uses. Exploitation of these properties of ZnO NPs in the protection of cultural heritage requires a deep knowledge of the rigors of this field and the adaptation of the novel nanomaterials to be compatible with the original materials of historical objects.

4. Application of ZnO nanoparticles in cultural heritage preservation

Cultural heritage monuments, although unanimously considered invaluable values of humanity, still in present days are exposed to various stress factors. Many of them cannot always be avoided and that inevitably result in the degradation and destruction of historic buildings or art objects. Thus, the development of novel techniques and materials with improved efficiency, that can be used in the cleaning and consolidation of art objects, is a very important issue in preventive conservation.

Due to their exceptional properties, ZnO nanoparticles are considered useful tools for development of multifunctional coatings with antibacterial, superhydorphobic, UV blocking, self-cleaning and anticorrosion properties. In the last decade, hybrid materials with ZnO nanoparticles embedded in various matrix have been used in conservation of historic buildings, paper artifact, textiles, etc.

4.1 ZnO nanoparticles - based composite materials in conservation of paper

Historic documents on paper are very sensitive, especially if they are kept in improper conditions, in spaces with high humidity, which favors the microbial attack. In addition, for documents with chromatic decorations or various types of ink, the pigments are subjected to a process of accentuated degradation under the conditions of lighting with strong UV radiation. The cellulose support itself can be damaged, with changes in strength and integrity due to prolonged exposure to sunlight, due to the action of UVA and UVB components. Jia et al. [43] propose a coating material for paper artifacts based on nanocellulose and ZnO nanoparticles. Hydroxypropyl cellulose (Klucel) was chosen because it is widely used in consolidating collections of documents, due to its very good compatibility with cellulose substrate. The ZnO nanoparticles were synthesized based on a facile hydrothermal reaction and obtained nanopowder was suspended in isopropanol together with pure nanocellulose to form a stabile dispersion. The treatment was applied on paper sample (Chinese newspaper from 1960) and proved to be efficient as antifungal agent against Aspergillus niger, Aspergillus versicolor and Rhizopus nigricans, common fungi that can be found in archives. Unexpectedly, the results obtained in this paper suggest similar antifungal activity regardless of the size of zinc oxide nanoparticles. The antibacterial and antifungal activity is dose dependent, the increase of ZnO concentration lead to an enhanced effect, but the composition should be adjust in order to not produce changes in visual appearance of the paper. In the study an optimal concentration of 0.55 g of ZnO nanoparticles in 100 mL Klucel solution is reported to inhibit the proliferation of fungi on treated samples without any visual effect on the paper.

The ability of ZnO nanopowders to absorb efficiently UV radiation has been exploited in recent decades for the production of industrial coatings for surfaces frequently exposed to sunlight or artificial light. Similar materials have been tested for the preventive conservation of historic paper artifacts, in order to protect them from the effect of prolonged lighting [44]. Functional coatings have been prepared from ZnO nanoparticles with average diameter 150 nm dispersed in ethanol (0.2 g% concentration) or ZnO NPs disperded in hydroxypropyl cellulose (Klucel) ethanolic solution, and deposited on the paper surface by a simple spraying procedure. Model paper artifacts, prepared by coloring samples with Alizarin using traditional techniques, were used to prove the protective effect of ZnO coating against dye degradation under UV exposure. Samples of untreated colored paper, paper coated with Klucel, paper coated with ZnO and paper treated with ZnO NPs in Klucel solution were subjected to irradiation with UV lamp (λ 270 nm, 30 W). Color fading (expressed as % of Alizarin absorbance reduction) after 120 minutes of exposure was 19% in the case of the paper treated with the nanocomposite ZnO NPs in Klucel, significantly lower compared to values obtained for untreated paper and paper treated with Klucel (34% and 31%, respectively). The coating with ZnO NPs in Hydroxipropyl cellulose (Klucel) deposited on sample of old paper possesses a high degree of transparency, thus produces no changes in color or gloss of the document (**Figure 3**).

The ability of nanoparticle –based coatings with ZnO and Ag to inhibit the growth of microorganism on the historic document was tested on strains isolated

Figure 3.
The visual aspect of old paper sample untreated (upper) and coated with nanocomposite ZnO NPs –Klucel (bottom) [44].

from a 1677 A.D. manuscript named "Serh Senk Adlky Geld Badr"("The biography of prophet Mohamed, Peace be upon him") deposited in Al-Azhar library, Cairo, Egypt [45]. Both Ag and ZnO NPs applied as treatment have been prepared using eco-friendly methods, biosynthesis mediated by endophytic fungal strain *Penicillium chrysogenum* and *Fusarium keratoplasticum*. The growth inhibition percentage of main strains identified on the studied documents (Bacillus subtilis and *Penicillium chrysogenum*) inoculated on model filter paper previously treated with Ag and ZnO NPs show very good performance. After 21 days, both Ag and ZnO NPs at low concentration of 1 M inhibit completely the development of *B. subtilis* colonies, while the *P. chrysogenum* were more resistant with an inhibition percentage of 59.9 produced by Ag and 51.9 by ZnO nanoparticles at the same concentration.

4.2 ZnO nanoparticles-based materials for the protection of stone heritage

Another interesting application of ZnO nanoparticles in development of functional coatings is devoted to consolidation and protective treatment of stone monuments or statues. An important number of heritage buildings are made from calcareous stone, very sensitive to the degradation by mineral transformation as a results of environmental stressors (temperature variation, humidity, rain, solar radiation, etc). Synergistic destruction effect occurs when the monument is exposed to microorganisms and their metabolic products. Many different microorganisms, such as algae, cyanobacteria, yeast, fungi, and lichens could colonize the surface or inner pores of stone, and their presence is a particular threat for monuments located in tropical climate, due to the high temperature and humidity conditions that favors microbial growth. The same antimicrobial and UV blocking properties of ZnO are used, in order to inhibit the colonization of stone surface with microorganism and to prevent the photo-oxidation of decoration due to the exposure to radiation. In addition, the ability of ZnO nanoparticles embedded in filmogenic matrices to produce suitable roughness leads to obtaining superhydrophobic coatings.

Fernandez et al. [46] reported the photocatalytic and antifungal properties of ZnO nanoparticles compared to MgO and Zn doped MgO NPs. Zinc doped nanoparticles exhibits the best performances both in photocatalytic destruction of methylene blue and in antimicrobial efficiency against many fungal species which produce stone deterioration, such *A. niger*, *P. oxalicum* and *P. maculans.*

Fungal infection is also responsible for biodeterioration of marble columns, thus ZnO antimicrobial properties recommend it for protective treatment of such monuments. Aldosari et al. [47] proposed a simple material obtained by dispersing commercial ZnO nanoparticles in acrylic polymer as a multifunctional biocidal and consolidating coating. The novel nanocomposite coating was applied to marble samples collected from various archeological sites in Egypt. The contact angle of water is enhanced from 112^0 on plain marble to 125^0 on marble treated with polymeric solution, and to a significantly higher value (140^0) in the case of treatment with ZnO-polymer nanocomposite. The color variation of marble sample subjected to various aging procedure prove the efficient protection of the treatment with ZnO NPs against UV exposure, together with minimal or no changes in optical properties following the coating.

Combination of ZnO nanorods and graphene nanoplatelets was proposed by Schifano et al. [48] as coating protective material for the prevention of biodeterioration produced by two bacteria (Arthrobacter aurescens and Achromobacter spanius) isolated from the Temple of Concordia (Agrigento's Valley of the Temples in Sicily, Italy). Three different materials Noto stone, Carrara marble and yellow bricks were tested as model lithic materials. The ZnO nanorods exhibit a high efficiency in reduction of bacterial growth for all strains (2% cell viability), at concentration ranging from 10 to 100 μg/mL. An unexpected result was obtain in antibiofilm properties tests. The biofilm formation on the samples treated with ZnO nanorods at concentration of 125 μg/mL was investigated. The ZnO nanomaterial is able to inhibit the adhesion of A. aurescens TC4 and A. spanius TC7 species on glass and plastic, while an effect of stimulation of biofilm formation was found for A. spanius TC1. Addition of graphene nanoplates to ZnO nanorods results in enhancement of the antibacterial activity and lead to biofilm inhibition also in the case of A. spanius TC1 strains.

One of the few cases of application of ZnO NPs coatings on real monuments is reported by van der Welf et al. [49]. Coating materials were prepared by mixing ZnO and Cu NPs with water-repellant or consolidant products commonly used in consolidation/restoration of historic stone monuments: Estel1000, Estel1100, and Silo111. This ready-to-use materials are based on siloxane oligomers (Silo111), tetraethoxysilane (Estel1000) and a combination of tetraethoxysilane and siloxane oligomers (Estel1100). Since the laboratory tests on similar calcareous stones from south Italy show that the treatment with 0.5w% ZnO NPs in commercial consolidant products Estel and Silo produced negligible changes in stone color and successfully inhibit the growth of *A. niger* fungal strains, in situ experiments were performed on an external area of the 12th-century church of San Leonardo di Siponto, Manfredonia, Italy (**Figure 4**).

As it is expected, the visual aspect after the application of ZnO –based nanocomposites is less modified compared to the use of Cu-NPs embedded in Estel1100 and Silo111. Chromatic variations produced by ZnO NPs were found to be below limit values after 6 months of application, and the biological colonization was reduced.

Figure 4.
Test area on the exterior of the san Leonardo di Siponto church (Manfredonia, Italy) (left). Effect of ZnO NPs treatment (right): A1 untreated area; A2 cleaned area; B1 cleaned area treated with Silo111; B2 cleaned area treated with Estel1100; C1 cleaned area treated with Silo111/ZnO-NPs; C2 cleaned area treated with Estel1100/ ZnO-NPs; D1 cleaned area treated with Silo111/Cu-NPs; D2 cleaned area treated with Estel1100/Cu-NPs [49].

4.3 ZnO nanoparticles-based material for painting protection

Materials with ZnO NPs have also proven effective in protecting paper artifacts decorated with oil painting. El-Feky et al. [50] reported a simple transparent coating for protection of oil paintings on paper against dirt accumulation, UV aging and fungal attack.

ZnO NPs prepared by a simple hydrothermal method, with size ranging from 122 to 286 nm were dispersed at 2% concentration in a commercial varnish to obtain the coating material.

The effect on the protecting dyes against UV radiation exposure, microbial colonization and dirt was tested using Fabriano paper sheets painted with linseed oil-based colors.

The dyes tested were famous coloring materials used in different period: vermilion (mercuric sulfide) for red color, yellow ochre (iron oxide) for yellow color and ultramarine blue. The painted paper samples were varnished using a commercial acrylic varnish with or without ZnO nanoparticles. The presence of ZnO NPs in varnish coating drastically reduced the deposition of dirt during the 6 month period of exposure, compared to unprotected or simply varnished colored paper (**Figure 5**). The cleaning procedure using a soft brush very easy for the areas protected with the varnish containing ZnO nanoparticles. The ZnO-based varnish proved to be effective to reduce the UV aging of oil painted paper, in particular for ultramarine blue, which is the most sensitive color.

4.4 ZnO nanoparticles-based material for metal conservation

Corrosion is a process that severely affects all types of industrial construction, thus reducing its effects is still a major problem, even for advanced materials used in the aeronautical industry, in electrical components, ships used in maritime transport, oil pipelines, etc. Significant efforts have been made in nanomaterial research to develop nanoparticle-based coatings with increased efficiency in metal protection. Zinc oxide NPs have proven to be a low cost choice, easy to synthesize and process into film-forming materials, and highly effective as an anticorrosive treatment for many types of metals.

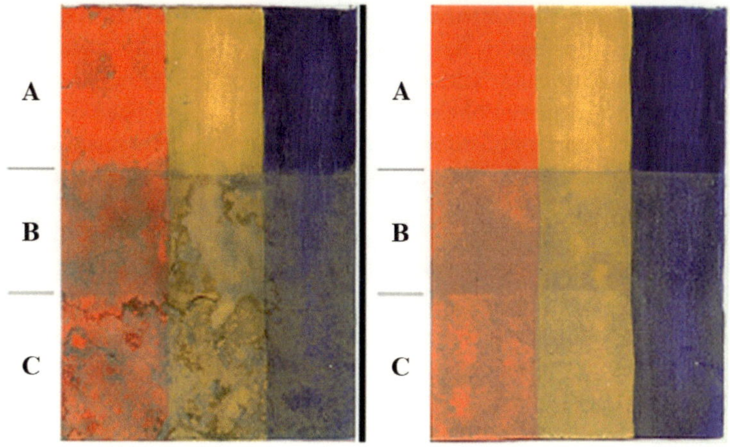

Figure 5.
Optical images of colored paintings after being left in open air for 6 months (left) and after cleaning with a brush and a dry piece of cotton (right). Area A: Painted with color + varnish + ZnO nanoparticles; area B: Painted with color + varnish; area C: Painted with color [50].

Composite materials consisting in ZnO NPs in various polymeric matrices exhibit superior performance from the point of view of corrosion resistance, since the presence of nanoparticles improve the properties of simple coatings in two aspects. The first effect is the enhancement of the barrier performance of the polymeric coating in the presence of ZnO nanopowder as filler, and the other one is the active anticorrosion action of ZnO material itself. Zinc oxide was extensively used in coating due to this active mechanism of corrosion inhibition, based to the possibility to slowly release Zn^{2+} cations through the solubility in water. Zinc ions react with hydroxyl groups on the metal surface (OH groups are present at cathodic corrosion sites on metals), with the formation of a protective layer preventing further reaction with the Fe^{2+} ions. Nanocoatings with various metal oxide nanoparticles (Al_2O_3, Fe_2O_3 and ZnO) at different concentrations in epoxy polymeric films were recently reported [51]. The performances in corrosion protection of carbon steel samples were tested, and the Bose plots show that even at low content of nanofiller in the polymeric matrix (i.e. 1 wt.%) the corrosion resistance of the film deposited on metal is significantly higher compared to plain epoxy treatment. At low concentrations of nanoparticles in the composite (1 wt.%) coating parameters expressed as coating resistance and non-ideal coating capacitance show similar values for all Al_2O_3, Fe_2O_3 and ZnO nanoparticles. Large differences are recorded at concentrations of 3 wt.%, where Al_2O nanoparticles prove to be more efficient than the others.

This type of coating, with low ZnO NPs content, could provide a solution as preventive anti-corrosion treatment for heritage buildings with steel elements, for example old skyscrapers in the historic New York City area or industrial heritage buildings in Manchester, England.

Purcar et al. [52] studied the influence of the silane derivative on the morphological, optical and wettability properties of hybrid nanomaterials based on ZnO nanoparticles. Organo-modified silane derivative (3-glycidyloxypropyl) trimethoxysilane (GPTMS), phenyl triethoxysilane (PhTES), octyltriethoxysilane (OTES), and octadecyltriethoxysilane (ODTES) have been used for the synthesis of coatings. The hybrid materials deposited on glass and metal leads to the formation of highly hydrophobic surfaces, with water contact angle up to 142^0. The hybrid coatings deposited on metallic coupons of Al, Cu and Fe exhibit good anticorrosion efficiency (unpublished data), in particular for the materials with octyltriethoxysilane (OTES), and octadecyltriethoxysilane (ODTES).

Despite the efforts made for the introduction of nanomaterials in the conservation of metallic heritage objects, nanoparticle – based treatments have failed to replace the classic solutions for anticorrosive protection [53]. Since the capability of nanocoatings, such as ZnO nanohybrids described above have been demonstrated in various cases on different metal alloy, with various organic or inorganic compatible matrices, further research is needed in order to develop and test materials designed for cultural heritage use.

4.5 ZnO nanoparticles-based material for textile conservation

The functionalization of textiles with coatings with various compositions, in order to ensure their antibacterial, anti-soiling and superhydrophobic properties is currently a trend in the modern textile industry. ZnO NP-based coatings are intensively studied and remarkable progress is reported for modification of various types of fabrics, from natural such as wool, cotton to synthetic ones [54].

As regarding the use of nanocoatings for the conservation of metal artifacts, case studies reporting the application of nanomaterials for the treatment of historic textiles are very rare.

In particular, the role of the prevention of the biological contamination of textile cultural heritage is to be taken into account. One of the very common causes of the damage of art and historical objects on textile support is the biodegradation produced by various microorganisms [55]. Although the antimicrobial and antifungal efficiency has been proven on a wide variety of microorganisms, in papers dealing with the application of functionalized fabrics for clothing or garments, no studies have been published on possibilities to use ZnO properties to inhibit the growth of specific fungi found on historic textiles under inadequate storage conditions.

The effect of ZnO nanoparticles on the fabrication of natural batik with antibacterial properties is reported by Eskany et al. [56]. Batik is a traditional Indonesian cloth, included since 2009 as part of Intangible Cultural Heritage of Humanity. The term batik refers to an artifact produced in a specific manner of coloring with natural dyes and wax the textile fabric, in order to obtain desired patterns. The most used support is cotton, which turns in a major disadvantage due to the overgrown with bacteria. ZnO NPs dispersions (1–2% concentration) have been applied on cotton textile samples before or after batik process, by using a common pad-dry-cure method. The antibacterial effect was tested on Gram-positive bacteria *S. aureus* and Gram-negative bacteria *E. coli*, with better results when ZnO was applied after the batik process. The ZnO treated batik cloth show good resistance to the microbial attack, together with an improvement of the quality of coloring.

5. Conclusions

Nanotechnology offers numerous solutions with increased efficiency both in cleaning and consolidation of artistic and historic artifacts, from microemulsion and gels to nanoparticles and nanocomposites. While other nanoparticles, such as silica or $Ca(OH)_2$ ones are already used in the protection of cultural heritage monuments, even they are not routine treatments, ZnO NPs still not gain their place in the practice of restoration of art works.

The fabrication of ZnO nanopowders for various industrial applications leads to the development of many methods, as they are summarized in this chapter. A large variety of size and morphology of the ZnO material could be obtained, resulting in variation of optical, electrical, photocatalytic and antimicrobial properties. In particular, UV blocking and antibacterial and antifungal activity of ZnO nanoparticles were discussed, since such properties could be employed in protection of the cultural objects against the main stress factors that induce degradation processes. The antimicrobial and antibiofilm activity of ZnO NPs against fungal and bacterial strains isolated from various archeological and historical sites have been proved, in addition to minor changes produced on visual aspect of the treated samples. The inhibition of microorganism growth was effective on various substrates (stones, marble, paper, textiles), thus the protective coatings containing ZnO NPs could be produced in a customable manner to fit many cultural heritage objectives.

One of the major barrier in extend the acceptability of ZnO-based treatments is the public concern on the environmental impact due to the release of the nanoparticles from the nanomaterials used in conservation/restoration procedure at large scale applied on monuments. Further research are needed to tailor the ZnO NPs to increase the efficiency and compatibility with the historic material, but also to assess their ecotoxicological long-term effect.

Acknowledgements

This work was supported by grants of the Romanian National Authority for Scientific Research and Innovation, CCCDI - UEFISCDI, project number PN-III-P1-1.2-PCCDI-2017-0743 and project number PN-III-P1-1.2-PCCDI-2017-0686, within PNCDI III.

Conflict of interest

The author(s) declared no potential conflicts of interest.

Author details

Ludmila Otilia Cinteză[1,2]* and Maria Antonia Tănase[1]

1 Physical Chemistry Department, University of Bucharest, Bucharest, Romania

2 ArchaeoScience#RO Department, Research Institute of the University of Bucharest (ICUB), Bucharest, Romania

*Address all correspondence to: ocinteza@gw-chimie.math.unibuc.ro

References

[1] Özgür Ü, Avrutin V, Morkoç H. Zinc oxide materials and devices grown by molecular beam Epitaxy. In: Molecular Beam Epitaxy. Elsevier; 2018. pp. 343-375. DOI: 10.1016/B978-0-12-387839-7.00016-6

[2] Shaban M, Mohamed F, Abdallah S. Production and characterization of Superhydrophobic and antibacterial coated fabrics utilizing ZnO Nanocatalyst. Scientific Reports. 2018;**8**(1):3925. DOI: 10.1038/s41598-018-22324-7

[3] Hatamie A, Khan A, Golabi M, Turner APF, Beni V, Mak WC, et al. Zinc oxide nanostructure-modified textile and its application to biosensing, Photocatalysis, and as antibacterial material. Langmuir. 2015;**31**(39):10913-10921. DOI: 10.1021/acs.langmuir.5b02341

[4] Kim K-B, Kim YW, Lim SK, Roh TH, Bang DY, Choi SM, et al. Risk assessment of zinc oxide, a cosmetic ingredient used as a UV filter of sunscreens. Journal of Toxicology and Environmental Health, Part B. 2017;**20**(3):155-182. DOI: 10.1080/10937404.2017.1290516

[5] Vidor FF, Meyers T, Müller K, Wirth GI, Hilleringmann U. Inverter circuits on freestanding flexible substrate using ZnO nanoparticles for cost-efficient electronics. Solid-State Electronics. 2017;**137**:16-21. DOI: 10.1016/j.sse.2017.07.011

[6] Gambino M, Ahmed MAA, Villa F, Cappitelli F. Zinc oxide nanoparticles hinder fungal biofilm development in an ancient Egyptian tomb. International Biodeterioration & Biodegradation. 2017;**122**:92-99. DOI: 10.1016/j.ibiod.2017.05.011

[7] Ismail AM, Menazea AA, Kabary HA, El-Sherbiny AE, Samy A. The influence of calcination temperature on structural and antimicrobial characteristics of zinc oxide nanoparticles synthesized by sol–gel method. Journal of Molecular Structure. 2019;**1196**:332-337. DOI: 10.1016/j.molstruc.2019.06.084

[8] Zhu L, Li Y, Zeng W. Hydrothermal synthesis of hierarchical flower-like ZnO nanostructure and its enhanced ethanol gas-sensing properties. Applied Surface Science. 2018;**427**:281-287. DOI: 10.1016/j.apsusc.2017.08.229

[9] Lin L, Peng X, Chen S, Zhang B, Feng Y. Preparation of diverse flower-like ZnO nanoaggregates for dye-sensitized solar cells. RSC Advances. 2015;**5**:25215-25221. DOI: 10.1039/C5RA01938A

[10] Ekthammathat N, Thongtem S, Thongtem T, Phuruangrat A. Characterization and antibacterial activity of nanostructured ZnO thin films synthesized through a hydrothermal method. Powder Technology. 2014;**254**:199-205. DOI: 10.1016/j.powtec.2014.01.010

[11] Stanković A, Dimitrijević S, Uskoković D. Influence of size scale and morphology on antibacterial properties of ZnO powders hydrothemally synthesized using different surface stabilizing agents. Colloids and Surfaces B: Biointerfaces. 2013;**102**:21-28. DOI: 10.1016/j.colsurfb.2012.07.033

[12] Kumaresan N, Ramamurthi K, Ramesh Babu R, Sethuraman K, Moorthy Babu S. Hydrothermally grown ZnO nanoparticles for effective photocatalytic activity. Applied Surface Science. 2017;**418**:138-146. DOI: 10.1016/j.apsusc.2016.12.231

[13] Senthilkumar N, Vivek E, Shankar M, Meena M, Vimalan M, Potheher IV. Synthesis of ZnO nanorods by one step microwave-assisted

hydrothermal route for electronic device applications. Journal of Materials Science: Materials in Electronics. 2018;**29**(4):2927-2938. DOI: 10.1007/s10854-017-8223-5

[14] Hasanpoor M, Aliofkhazraei M, Delavari H. Microwave-assisted synthesis of zinc oxide nanoparticles. Procedia Materials Science. 2015;**11**: 320-325. DOI: 10.1016/j.mspro.2015.11.101

[15] Zare M, Namratha K, Byrappa K, Surendra DM, Yallappa S, Hungund B. Surfactant assisted solvothermal synthesis of ZnO nanoparticles and study of their antimicrobial and antioxidant properties. Journal of Materials Science & Technology. 2018;**34**(6):1035-1043. DOI: 10.1016/j.jmst.2017.09.014

[16] Krishnapriya R, Praneetha S, Vadivel Murugan A. Energy-efficient, microwave-assisted hydro/solvothermal synthesis of hierarchical flowers and rice grain-like ZnO nanocrystals as photoanodes for high performance dye-sensitized solar cells. CrystEngComm. 2015;**17**(43):8353-8367. DOI: 10.1039/C5CE01438G

[17] Cinteza LO, Purcar V, Gifu C, Ianchis R, Petcu C. Supercritical CO2 assisted synthesis of flower – like ZnO nanoparticles. Journal of Optoelectronics and Advanced Materials. 2017;**19**(11-12):800-805

[18] Agarwal H, Venkat Kumar S, Rajeshkumar S. A review on green synthesis of zinc oxide nanoparticles – An eco-friendly approach. Resource-Efficient Technologies. 2017;**3**(4):406-413. DOI: 10.1016/j.reffit.2017.03.002

[19] Fakhari S, Jamzad M, Kabiri Fard H. Green synthesis of zinc oxide nanoparticles: A comparison. Green Chemistry Letters and Reviews. 2019;**12**(1):19-24. DOI: 10.1080/17518253.2018.1547925

[20] Hassan Basri H, Talib RA, Sukor R, Othman SH, Ariffin H. Effect of synthesis temperature on the size of ZnO nanoparticles derived from pineapple Peel extract and antibacterial activity of ZnO–starch Nanocomposite films. Nanomaterials. 2020;**10**(6):1061. DOI: 10.3390/nano10061061

[21] Aminuzzaman M, Ying LP, Goh W-S, Watanabe A. Green synthesis of zinc oxide nanoparticles using aqueous extract of Garcinia mangostana fruit pericarp and their photocatalytic activity. Bulletin of Materials Science. 2018;**41**:50. DOI: 10.1007/s12034-018-1568-4

[22] Selim YA, Azb MA, Ragab I, Abd El HM, Azim M. Green synthesis of zinc oxide nanoparticles using aqueous extract of Deverra tortuosa and their cytotoxic activities. Scientific Reports. 2020;**10**(1):3445. DOI: 10.1038/s41598-020-60541-1

[23] Gawade VV, Gavade NL, Shinde HM, Babar SB, Kadam AN, Garadkar KM. Green synthesis of ZnO nanoparticles by using Calotropis procera leaves for the photodegradation of methyl orange. Journal of Materials Science: Materials in Electronics. 2017;**28**(18):14033-14039. DOI: 10.1007/s10854-017-7254-2

[24] Alavi M, Karimi N, Valadbcigi T. Antibacterial, Antibiofilm, Antiquorum sensing, Antimotility, and antioxidant activities of green fabricated Ag, Cu, TiO_2, ZnO, and Fe_3O_4 NPs via *Protoparmeliopsis muralis* lichen aqueous extract against multi-drug-resistant bacteria. ACS Biomaterials Science & Engineering. 2019;**5**(9):4228-4243. DOI: 10.1021/acsbiomaterials.9b00274

[25] Suresh J, Pradheesh G, Alexramani V, Sundrarajan M, Hong SI. Green synthesis and characterization of zinc oxide nanoparticle using insulin plant (*Costus pictus D. Don*) and investigation of its antimicrobial as

well as anticancer activities. Advances in Natural Sciences: Nanoscience and Nanotechnology. 2018;**9**(1):015008. DOI: 10.1088/2043-6254/aaa6f1

[26] Theophil Anand G, Renuka D, Ramesh R, Anandaraj L, John Sundaram S, Ramalingam G, et al. Green synthesis of ZnO nanoparticle using Prunus dulcis (almond gum) for antimicrobial and supercapacitor applications. Surfaces and Interfaces. 2019;**17**:100376. DOI: 10.1016/j.surfin.2019.100376

[27] Doan Thi TU, Nguyen TT, Thi YD, Ta Thi KH, Phan BT, Pham KN. Green synthesis of ZnO nanoparticles using orange fruit peel extract for antibacterial activities. RSC Advances. 2020;**10**(40):23899-23907. DOI: 10.1039/D0RA04926C

[28] Rambabu K, Bharath G, Banat F, Show PL. Green synthesis of zinc oxide nanoparticles using Phoenix dactylifera waste as bioreductant for effective dye degradation and antibacterial performance in wastewater treatment. Journal of Hazardous Materials. 2021;**402**:123560. DOI: 10.1016/j.jhazmat.2020.123560

[29] Ogunyemi SO, Abdallah Y, Zhang M, Fouad H, Hong X, Ibrahim E, et al. Green synthesis of zinc oxide nanoparticles using different plant extracts and their antibacterial activity against *Xanthomonas oryzae* pv. Oryzae. Artificial Cells, Nanomedicine, and Biotechnology. 2019;**47**(1):341-352. DOI: 10.1080/21691401.2018.1557671

[30] Hossain A, Abdallah Y, Ali MA, Masum MMI, Li B, Sun G, et al. Lemon-fruit-based green synthesis of zinc oxide nanoparticles and titanium dioxide nanoparticles against soft rot bacterial pathogen Dickeya dadantii. Biomolecules. 2019;**9**(12):863. DOI: 10.3390/biom9120863

[31] Abdelkhalek A, Al-Askar AA. Green synthesized ZnO nanoparticles mediated by Mentha Spicata extract induce plant systemic resistance against tobacco mosaic virus. Applied Sciences. 2020;**10**(15):5054. DOI: 10.3390/app10155054

[32] Sorbiun M, Shayegan Mehr E, Ramazani A, Taghavi Fardood S. Green synthesis of zinc oxide and copper oxide nanoparticles using aqueous extract of oak fruit Hull (Jaft) and comparing their Photocatalytic degradation of basic violet 3. Int J Environ Res. 2018;**12**(1):29-37. DOI: 10.1007/s41742-018-0064-4

[33] Sharmila G, Thirumarimurugan M, Muthukumaran C. Green synthesis of ZnO nanoparticles using Tecoma castanifolia leaf extract: Characterization and evaluation of its antioxidant, bactericidal and anticancer activities. Microchemical Journal. 2019;**145**:578-587. DOI: 10.1016/j.microc.2018.11.022

[34] Kumar R, Umar A, Kumar G, Nalwa HS. Antimicrobial properties of ZnO nanomaterials: A review. Ceramics International. 2017;**43**(5):3940-3961. DOI: 10.1016/j.ceramint.2016.12.062

[35] de Lucas-Gil E, Leret P, Monte-Serrano M, Reinosa JJ, Enríquez E, Del Campo A, et al. ZnO Nanoporous spheres with broad-Spectrum antimicrobial activity by physicochemical interactions. ACS Appl Nano Mater. 2018;**1**(7):3214-3225. DOI: 10.1021/acsanm.8b00402

[36] Raghupathi KR, Koodali RT, Manna AC. Size-dependent bacterial growth inhibition and mechanism of antibacterial activity of zinc oxide nanoparticles. Langmuir. 2011;**27**(7):4020-4028. DOI: 10.1021/la104825u

[37] El.Shafei A, Abou-Okeil A. ZnO/carboxymethyl chitosan bionano-composite to impart antibacterial and UV protection for cotton

fabric. Carbohydrate Polymers. 2011;**83**(2):920-925. DOI: 10.1016/j.carbpol.2010.08.083

[38] Lee M, Kwak G, Yong K. Wettability control of ZnO nanoparticles for universal applications. ACS Applied Materials & Interfaces. 2011;**3**(9):3350-3356. DOI: 10.1021/am2004762

[39] Ghasemi N, Seyfi J, Asadollahzadeh MJ. Imparting superhydrophobic and antibacterial properties onto the cotton fabrics: Synergistic effect of zinc oxide nanoparticles and octadecanethiol. Cellulose. 2018;**25**(7):4211-4222. DOI: 10.1007/s10570-018-1837-9

[40] Becheri A, Dürr M, Lo Nostro P, Baglioni P. Synthesis and characterization of zinc oxide nanoparticles: Application to textiles as UV-absorbers. Journal of Nanoparticle Research. 2008;**10**(4):679-689. DOI: 10.1007/s11051-007-9318-3

[41] Kumar SG, Rao KSRK. Zinc oxide based photocatalysis: Tailoring surface-bulk structure and related interfacial charge carrier dynamics for better environmental applications. RSC Advances. 2015;**5**(5):3306-3351

[42] Qing Y, Yang C, Hu C, Zheng Y, Liu C. A facile method to prepare superhydrophobic fluorinated polysiloxane/ZnO nanocomposite coatings with corrosion resistance. Applied Surface Science. 2015;**326**:48-54. DOI: 10.1016/j.apsusc.2014.11.100

[43] Jia M, Zhang X, Weng J, Zhang J, Zhang M. Protective coating of paper works: ZnO/cellulose nanocrystal composites and analytical characterization. Journal of Cultural Heritage. 2019;**38**:64-74. DOI: 10.1016/j.culher.2019.02.006

[44] Afsharpour M, Imani S. Preventive protection of paper works by using nanocomposite coating of zinc oxide. Journal of Cultural Heritage. 2017;**25**:142-148. DOI: 10.1016/j.culher.2016.12.007

[45] Fouda A, Abdel-Maksoud G, Abdel-Rahman MA, Salem SS, Hassan SE-D, El-Sadany MA-H. Eco-friendly approach utilizing green synthesized nanoparticles for paper conservation against microbes involved in biodeterioration of archaeological manuscript. International Biodeterioration and Biodegradation. 2019;**142**:160-169. DOI: 10.1016/j.ibiod.2019.05.012

[46] Sierra-Fernandez A, De la Rosa-García SC, Gomez-Villalba LS, Gómez-Cornelio S, Rabanal ME, Fort R, et al. Synthesis, Photocatalytic, and antifungal properties of MgO, ZnO and Zn/Mg oxide nanoparticles for the protection of calcareous stone heritage. ACS Applied Materials & Interfaces. 2017;**9**(29):24873-24886. DOI: 10.1021/acsami.7b06130

[47] Aldosari MA, Darwish SS, Adam MA, Elmarzugi NA, Ahmed SM. Using ZnO nanoparticles in fungal inhibition and self-protection of exposed marble columns in historic sites. Archaeological and Anthropological Sciences. 2019;**11**(7):3407-3422. DOI: 10.1007/s12520-018-0762-z

[48] Schifano E, Cavallini D, De Bellis G, Bracciale MP, Felici AC, Santarelli ML, et al. Antibacterial effect of zinc oxide-based Nanomaterials on environmental Biodeteriogens affecting historical buildings. Nanomaterials. 2020;**10**(2):335. DOI: 10.3390/nano10020335

[49] van der Werf ID, Ditaranto N, Picca RA, Sportelli MC, Sabbatini L. Development of a novel conservation treatment of stone monuments with bioactive nanocomposites. Herit Sci. 2015;**3**(1):29. DOI: 10.1186/s40494-015-0060-3

[50] El-Feky OM, Hassan EA, Fadel SM, Hassan ML. Use of ZnO nanoparticles for protecting oil paintings on paper support against dirt, fungal attack, and UV aging. Journal of Cultural Heritage. 2014;**15**(2):165-172. DOI: 10.1016/j.culher.2013.01.012

[51] Torknezhad Y, Khosravi M, Assefi M. Corrosion protection performance of nanoparticle incorporated epoxy paint assessed by linear polarization and electrochemical impedance spectroscopy. Materials and Corrosion. 2018;**69**(4):472-480. DOI: 10.1002/maco.201709563

[52] Purcar V, Şomoghi R, Niţu S, Nicolae C-A, Alexandrescu E, Gîfu I, et al. The effect of different coupling agents on Nano-ZnO materials obtained via the sol–gel process. Nanomaterials. 2017;**7**(12):439. DOI: 10.3390/nano7120439

[53] Artesani A, Di Turo F, Zucchelli M, Traviglia A. Recent advances in protective coatings for cultural heritage–an overview. Coatings. 2020;**10**(3):217. DOI: 10.3390/coatings10030217

[54] Verbič A, Gorjanc M, Simončič B. Zinc oxide for functional textile coatings: Recent advances. Coatings. 2019;**9**(9):550. DOI: 10.3390/coatings9090550

[55] Castrillón E, Rivera L, Palma Ramos AI, Castañeda Sánchez J, Elisa Drago Serrano M. Origin and control strategies of biofilms in the cultural heritage. In: Kırmusaoğlu S, editor. Antimicrobials, Antibiotic Resistance. IntechOpen: Antibiofilm Strategies and Activity Methods; 2019. DOI: 10.5772/intechopen.79617

[56] Eskani IN, Haerudin A, Setiawan J, Lestari DW, Isnaini AW. Modification of cotton fabric by ZnO nanoparticles for producing an antibacterial natural dyed batik. IOP Conf Ser Earth Environ Sci. 2020;**462**:012031. DOI: 10.1088/1755-1315/462/1/012031

Chapter 7

Deposition of Silver Nanoparticles on Indium Tin Oxide Substrates by Plasma-Assisted Hot-Filament Evaporation

Abtisam Hasan Hamood Al-Masoodi, Boon Tong Goh, Ahmed H.H. Al-Masoodi and Wan Haliza Binti Abd Majid

Abstract

Nanoparticles of noble metals have unique properties including large surface energies, surface plasmon excitation, quantum confinement effect, and high electron accumulation. Among these nanoparticles, silver (Ag) nanoparticles have strong responses in visible light region due to its high plasmon excitation. These unique properties depend on the size, shape, interparticle separation and surrounded medium of Ag nanoparticles. Indium tin oxide (ITO) is widely used as an electrode for flat panel devices in such as electronic, optoelectronic and sensing applications. Nowadays, Ag nanoparticles were deposited on ITO to improve their optical and electrical properties. Plasma-assisted hot-filament evaporation (PAHFE) technique produced high-density of crystalline Ag nanoparticles with controlling in the size and distribution on ITO surface. In this chapter, we will discuss about the PAHFE technique for the deposition of Ag nanoparticles on ITO and influences of the experimental parameters on the physical and optical properties, and electronic structure of the deposited Ag nanoparticles on ITO.

Keywords: silver nanoparticles, plasma-assisted hot-filament evaporation, properties, indium tin oxide, electronic structure

1. Introduction

Noble metallic nanoparticles, which are described as metals in the nanoscale with dimensions within size range from 1 to 100 nm, recently received significant attention in optoelectronic, biosensing and photocatalysts applications [1–4]. This is due to their unique properties compared to the bulk materials such as large surface energies, surface plasmon excitation, quantum confinement effect, and high electron accumulation. The bulk material has constant physical properties regardless of their size and shape, however, these properties of the nanoparticles are a function of their size, shape, distribution and surrounded medium. Among these nanoparticles, silver (Ag) nanoparticles have particularly strong responses in the visible light region due to its high plasmon excitation at threshold energy of around 3.9 eV (318 nm). A specific phenomenon of the nanoparticles is localized surface plasmon resonance (LSPR) which results from the collective oscillations of the free electrons on the metallic nanoparticle surfaces. Thus, the LSPR of Ag nanoparticles can be tuned to any wavelength in the visible light region. This is a highly desirable characteristic enabling the usage of Ag nanoparticles in optoelectronic devices mainly in solar cell and light emitting diode devices [5, 6]. The LSPR wavelength position can be tuned by varying the size, shape, particle spacing and compositions of the nanoparticles as well as a surrounding environment such as an insulating surface or presence of a dielectric layer [7]. Oxide layer can be form around the nanoparticles and acts as dielectric substances leading to formation of metal–metal oxide core-shell nanoparticles. These core-shell nanoparticles have been reported to produce wide SPR bands compered to pure metallic nanoparticle [8, 9]. Thus, the wide range of the LSPR existing through the metal oxide layer could be better than increasing in the nanoparticles size that may significantly lead to reduction in light scattering. On the other hand, indium tin oxide (ITO) is a transparent conducting oxide that has high transparency in visible light regions, low sheet resistance, and high work function. Moreover, ITO is widely used as anode material for optoelectronic devices as a hole injection layer in the devices [10, 11]. Thus, deposition of Ag nanoparticles layer on ITO suggests a feasible approach to enhance the flexibility, luminescent efficiency, electrical conductivity, and adhesion to device layers.

Ag nanoparticles layer are widely synthesized using evaporation-condensation, electron beam irradiation, and radio frequency plasma-assisted thermal evaporation, which show a good surface adhesion with the dielectric surface [12–14]. However, these physical deposition methods generally involve complicated structures, surface treatments, and high reaction temperatures up to several thousand °C in a plasma jet and 400°C for thermal annealing purposes [13, 15, 16]. The high reaction temperatures usually lead to the destruction of the device layers during the deposition process [13, 15, 16]. This issue can be avoided using thermal evaporation by hot-filament, as it provides a rapid evaporation of metallic nanoparticles source in high purity and the most important is it involving low substrate temperatures (usually below 400°C) [17, 18]. Moreover, plasma-driven deposition controls the transportation and deposition of the evaporated metallic adatoms, which directly leads to better size and uniformity of the deposition [19]. Nevertheless, pre-plasma treatment on the substrate surface proved to be more effective in removing organic contaminations that impede the particle mobility on the deposited surface [20, 21]. Thus, plasma-assisted hot filament evaporation technique is expected to deposition of Ag nanoparticles layer in uniform size and distribution at low substrate temperatures.

2. Plasma-assisted hot-filament evaporation

Evaporation is a common method for deposition of thin film from their source materials in a vacuum as a physical vapor deposition (PVD) technique. The source materials are evaporated using evaporation source such as metal boat or coiled wire. Tungsten is a metal and has very high melting temperature about 3422°C. A tungsten wire can be coiled to form spiral shape for using as a hot-filament. This hot-filament is preferred to use for deposition very thin film compared to a tungsten boat. However, the deposition rate of the thin film on substrates using PVD technique is very low. Other common technique, chemical vapor deposition (CVD) is a technique for deposition nanostructured thin film on substrates with very high temperature using precursor gases [22]. Insertion of hot-filament in to the CVD technique helps to deposit nanocrystalline of nanostructured thin film at lower substrate temperature as hot-filament chemical vapor deposition (HFCVD) technique [18, 23–27]. The HFCVD process employs the heated filament to decompose the precursor species and deposit nanostructured film on the substrate. On other hand, plasma can include electron, ions, free radicals, photons and neutrals that can generate reactive chemical species for enhancing the thin film deposition. The most recent technique was used is plasma-enhanced chemical vapor deposition (PECVD) [28–31]. However, PECVD requires a long time deposition and produces non-purity thin film with existence of toxic and explosive gases in plasma stream [32]. To avoid the existence of the toxic gases and contamination on the film, hydrogen plasma is used to remove the contaminations and provide purity thin film deposition without any toxic gases production. Therefore, hybrid of hot-filament with hydrogen plasma as plasma-assisted hot-filament evaporation (PAHFE) is a promise technique for deposition of purity metal nanoparticles layer on the substrate at low substrate temperature. Thus, the features of this technique are: (1) control of particle size, shape and interparticle separation, (2) enhancement of the nanoparticles crystallity, (3) stabilization in the physical and structure properties, (4) high deposition rate, (5) production of purity nanoparticles, (6) avoid particles aggregation, (7) low substrate temperature, and (8) fast deposition rate.

2.1 Structure of PAHFE

Figure 1(a) shows the real picture of the PAHFE system. The PAHFE consists to three main parts; reaction chamber, vacuum pumping system and supply units. The schematic diagram of the home-built of the reaction chamber is shown in **Figure 1(b)**. A radio frequency (RF) electrode was used as a plasma generation source. The tungsten wire with a diameter of 1 mm can be coiled to approximately 30 coils, generating coils with a diameter of 2 mm and length of 3 cm. This formed a tungsten coil filament as the hot-filament which the Ag source (Ag wire) was hanged on the filament coils for evaporation purpose. This hot-filament was placed below the RF electrode at a distance of 2 cm. Two copper electrodes were used to hang the tungsten wire, as shown in **Figure 1(b)**. Substrates which the nanoparticles deposited on were put on a substrate holder. The substrate holder was placed below the filament at a distance of 10 cm. After that, the substrates can be heated using a heater rod inserted into the substrate

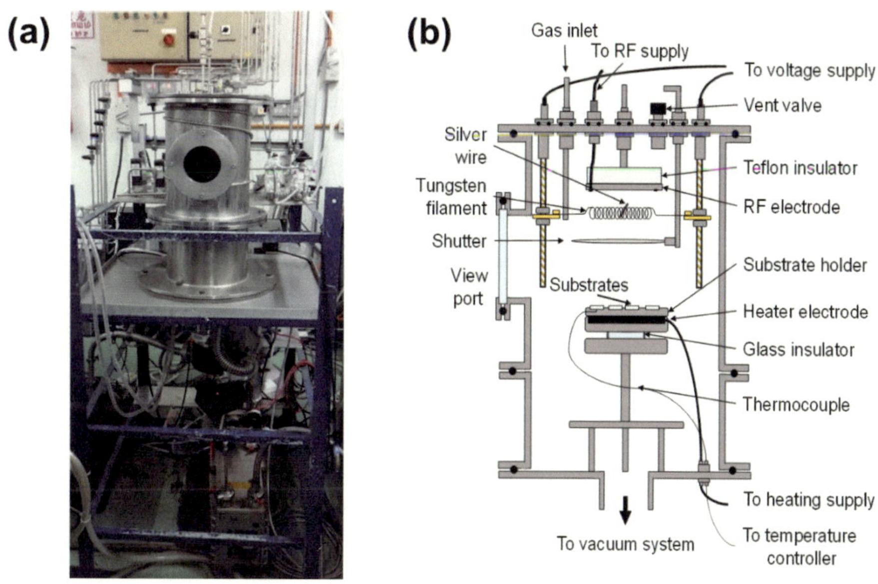

Figure 1.
(a) Real picture of PAHFE system. (b) Schematic diagram of the home-built PAHFE chamber [33].

holder for achieving a desired substrate temperature. A thermocouple was used to measure the substrate temperature and connected to a temperature controller. Finally, in order to control the evaporation time, a shutter was placed between the filament and substrate holder. There are four supply units for PAHFE technique. RF power supply of 13.56 MHz was connected to the RF electrode by a matching impedance for plasma generation in the reaction chamber. Filament heating power supply is used for heating the filament to a desired temperature in purpose of Ag wire evaporation. The two copper electrodes were connected to the filament heating power supply. Substrate heating power supply was connected to the heater rod to heat the substrate for achieving a particular substrate temperature. Finally, a hydrogen gas was supplied to the reaction chamber through ¼ inch SS tube and ball valve. The gas flow is precisely controlled by a mass flow-controller. Furthermore, the vacuum pumping system is used to evacuate the reaction chamber before and after deposition processes, and control the pressure as well.

2.2 Processes of PAHFE

Before placing the substrates inside the chamber, the substrates were put inside a beaker including Decon 90 diluted in deionized water to clean using ultrasound at 60°C for 60 minutes. Then, rinsing the substrates was sequentially done using deionized water, acetone and isopropyl alcohol. Finally, the clean substrates were dried using nitrogen gas. The Ag wire with a length of 2 mm and a diameter of 0.5 mm (Ag wire weight of 4.2 ± 0.2 mg and purity of 99.9%) was placed inside the coiled part of the filament. After placing the substrates inside the chamber, the chamber was evacuated to minimum or lowest pressure of around 5×10^{-3} Pa, using turbomolecular pump. Then, the substrates were heated to the desired substrate temperature. Before the nanoparticles deposition, plasma cleaning process was done on the substrate surface and Ag wire using hydrogen plasma. The parameters of the plasma cleaning process were put at 6 W, 100 sccm, 75 Pa, and 10 minutes for the RF power, hydrogen flow-rate, pressure, and time, respectively. In general, the hydrogen plasma process is utilized to remove any surface contaminants including native oxide on the substrates. After the plasma cleaning, the filament temperature was slowly increased to reach to 1600°C under hydrogen plasma ambient condition. The heated filament under plasma has sufficient thermal energy to completely evaporate the Ag wire which has a melting point of about 961°C. The deposition was started by opening the shutter subsequent to the plasma cleaning process as plasma deposition process. The RF power, hydrogen flow-rate, pressure, and time were fixed at 6 W, 50 sccm, 41 Pa, and 3 minutes, respectively. **Figure 2** shows the

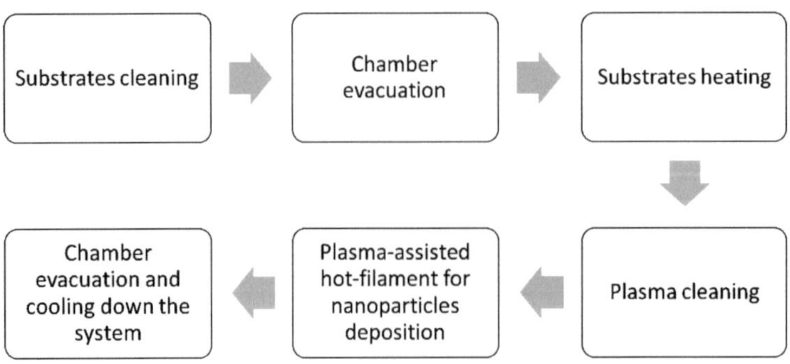

Figure 2.
Scheme of the PAHFE processes.

scheme of the PAHFE processes. The substrate temperature was varied between 25°C and 250°C at 25°C, 80°C, 140°C, 200°C, and 250°C.

3. Salient features of the deposited Ag nanoparticles on ITO substrate

The properties of Ag material in the nanoparticles form are different from the Ag bulk material due to consisting of the nanoparticles from down to few atoms. Thus, these few atoms may dominate of the nanoparticles properties. Also, distribution of the nanoparticles on a layer may strongly affect on their properties. This distribution of Ag nanoparticles on the layer depends on their size and interparticle separation. So, the nanoparticles could stand as a two-dimensional array to form Ag nanoparticles layer. Moreover, it is well-known LSPR characteristics of Ag nanoparticles that have unique optical properties in all visible region. This LSPR of Ag nanoparticles strongly depends on the physical properties of the Ag nanoparticles layer. Furthermore, PAHFE technique can control the properties of Ag nanoparticles layer by variation in growth parameters. Thus, the changing in these properties of the nanoparticles layer via the growth parameters is discussed, as mentioned below.

3.1 Morphological properties

The surface morphology including size, shape and distribution of Ag nanoparticles on the ITO/glass substrates, which can be obtained using field emission scanning electron microscopy (FESEM), varies with variation of the experimental parameters such as growth environment and substrate temperatures. FESEM images of the deposited Ag nanoparticles on ITO substrates at substrate temperature of 25°C without and with using plasma conditions during the deposition processes are presented in **Figure 3(a and b)**. These nanoparticles showed spherical shape in both conditions, and inconsistency size and interparticle separation distributions at no plasma condition compared with using plasma, as shown in

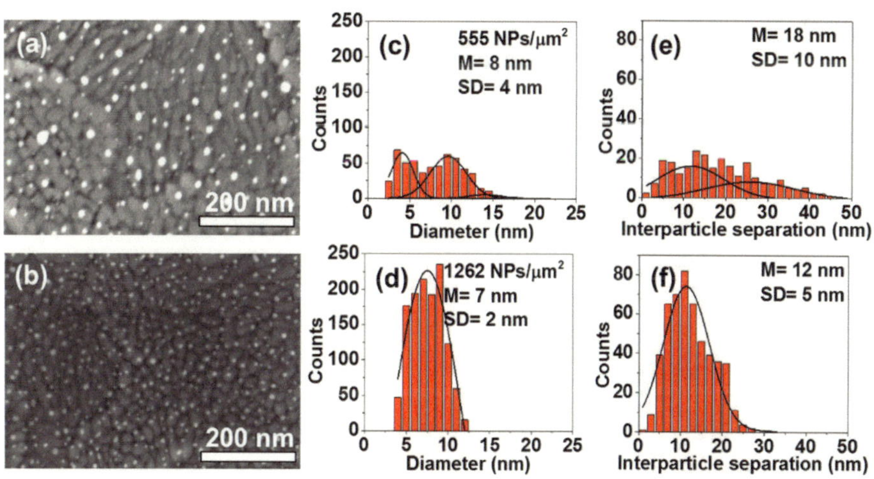

Figure 3.
(a and b) FESEM images, (c and d) histograms of diameter, and (e and f) histograms of interparticle separation of Ag nanoparticles deposited on ITO/glass substrates without and with using plasma at room temperature of the substrate temperature. The black solid lines in the histograms are the fitting Gaussian curves, and the nanoparticle density in NPs/μm² and the mean diameters or interparticle separations (M) with standard deviation (SD) values are inserted in each sub-figure.

Figure 3(c–f). Density of the nanoparticles at with plasma condition is about 1262 nanoparticles/μm^2 and showed higher than the density at without plasma condition (555 nanoparticles/μm^2). Thus, the combination of the hot-filament temperature and hydrogen ion plasma during the deposition assists to facilitate the growth of high density of Ag nanoparticles in uniform size and interparticle spacing. Moreover, the hot-filament temperature of 1600°C provided the evaporated Ag adatoms kinetic energy at low deposition pressure of 41 Pa. This thermal induced kinetic energy is sufficient to diffuse the Ag adatoms on the ITO surface for fast-forming of Ag nanoparticles at room temperature [34].

For variation in the substrate temperatures of 80, 140, 200 and 250°C with using plasma during the cleaning and deposition processes, the FESEM images of the deposited Ag nanoparticles on ITO substrates are presented in **Figure 4(a–d)**. In the substrate temperatures up to 140°C, Ag nanoparticles appeared to be also in spherical shape with high consistency in size and distribution. Also, the nanoparticles size increases with increasing in the temperature up to 140°C. This method is directly deposit the nanoparticles on surface in uniform size and interparticle separation without any additional process such as annealing process. This is in contrast

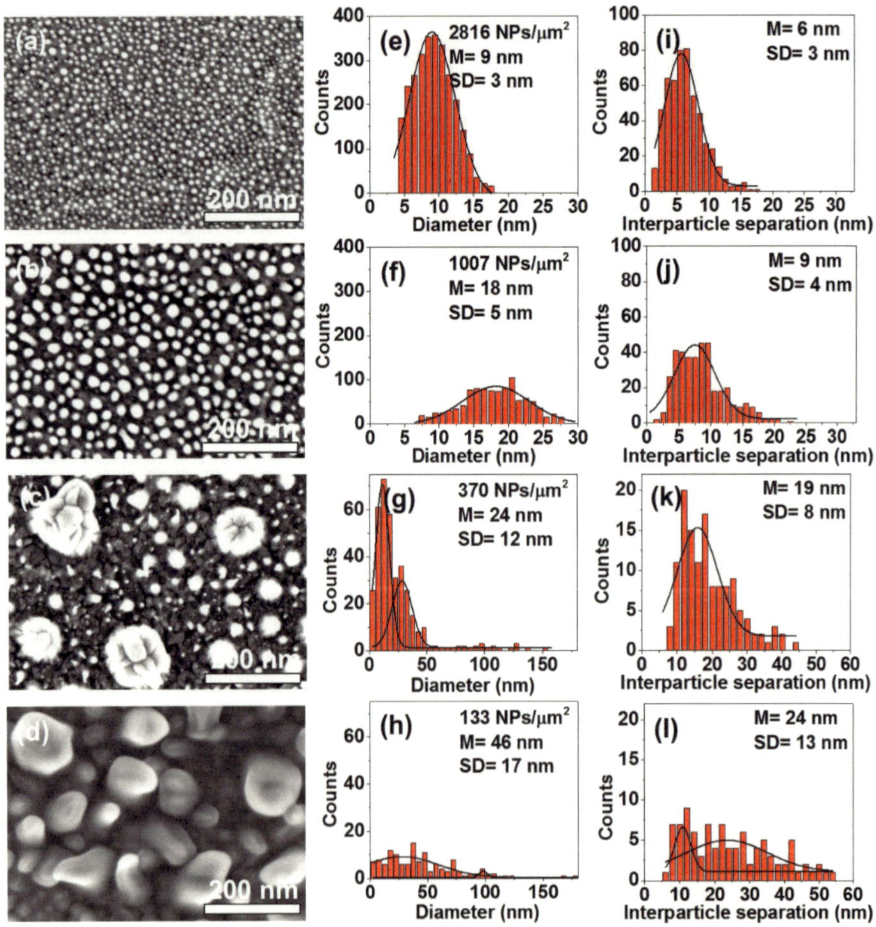

Figure 4.
(a–d) FESEM images, (e–h) histograms of diameter, (i–l) histograms of interparticle separation of Ag nanoparticles deposited on ITO/glass substrates using plasma with different substrate temperatures of 80°C, 140°C, 200°C, and 250°C, respectively. The black solid lines in the histograms are the fitting Gaussian curves, and the nanoparticle density in NPs/μm^2 and the mean diameters or interparticle separations (M) with standard deviation (SD) values are inserted in each sub-figure.

to the other nanoparticles formation methods obtained by other research studies which illustrated the nanoparticles formation of different sizes and interparticle separation as the temperature increased. Also, those methods required subsequent process of annealing temperature as an additional process that showed indirect deposition on the substrate surface [35–38]. Moreover, the substrate temperature provides sufficient surface mobility to the deposited Ag nanoparticles for diffusion into their growth sites and coalescing to form larger particles [39]. In addition, the ionized hydrogen radicals in the plasma deposition can transfer their kinetic energies to the ITO surface leading to increase the surface diffusion of Ag adatoms that enhances the uniform distribution of the deposited Ag nanoparticles on the ITO surface. Density of the deposited Ag nanoparticles decreases with increase in the substrate temperature, with the highest density of Ag nanoparticles obtained at 80°C compared to the room temperature (**Figures 3(d)** and **4(e)**). This indicates that the substrate temperature of ITO/glass provides sufficient surface mobility for the Ag adatoms to form Ag nanoparticles with the highest density at 80°C. Moreover, the substrate temperature at 200°C and 250°C, the size of the nanoparticles increased in a non-uniformity and an inconsistent interparticle distribution, as seen in **Figure 4(c** and **d)**. The hydrogen plasma can etch out oxygen bonded to the ITO surface leading to formation of SnO and In_2O_3 particles at higher substrate temperatures more than the melting temperatures of In (156.6°C) and Sn (231.9°C). Then, the oxygen adatoms diffused to the Ag nanoparticles resulting in formation of Ag oxide layer [40]. These oxide layer and SnO and In_2O_3 particles appeared the nanoparticles in non-uniformity in the size and shape (**Figure 4(c** and **d)**). The area filling fraction is represented by the ratio of the nanoparticles area to the substrate area [41]. The calculated area filling fractions for the nanoparticles were 20.79, 26.96, 32.35, and 34.65% at substrate temperatures of 80°C, 140°C, 200°C, and 250°C, respectively. The increase in the area filling fraction of Ag nanoparticles with decreasing in the density as the substrate temperature increases on ITO/glass substrate is due to the increase in the size with decrease in interparticle separation, as shown in **Figure 4(e–l)**. Therefore, the physical properties of Ag nanoparticles on ITO substrates can be simply controlled by fine-tuning in the substrate temperature to below 200°C with using plasma during the growth processes.

3.2 Structural properties

The deposition of Ag nanoparticles on the oxide substrate under hydrogen plasma may affect on the composition of the nanoparticles. The nanoparticles structure can be recorded via X-ray diffraction (XRD) patterns. **Figure 5** shows the XRD patterns of the Ag nanoparticles deposited on ITO/glass substrates at different substrate temperatures. ITO diffraction peaks are obviously presented in prepared sample at 25°C with diffraction peaks shown in **Figure 5**, according to JCPDS card no. 01-089-4596. These diffraction peaks of ITO significantly decrease with increasing in the temperature due to the plasma etching on the ITO surface. A small noticeable diffraction peak at 32.8°, which corresponds to the Ag_2O crystalline plane of (111) according to JCPDS card no. 41-1104, appeared at all samples and increases with increasing in the temperature. An additional diffraction peak of Ag_2O at 68.8° for the crystalline plane of (222) can be observed at temperatures above 200°C. Moreover, AgO diffraction peaks are also observed in the prepared samples at all substrate temperatures above 80°C. These diffraction peaks of AgO located at 39.4, 56.88, and 67.3° correspond to the crystalline planes of (−202), (−113), and (−313), according to JCPDS card no. 022-0472. Moreover, very small diffraction peaks of metallic Ag appeared at 38.1 and 44.3° indicated to crystallographic planes of (111) and (200), respectively. The emergence of AgO and Ag_2O peaks is due to the

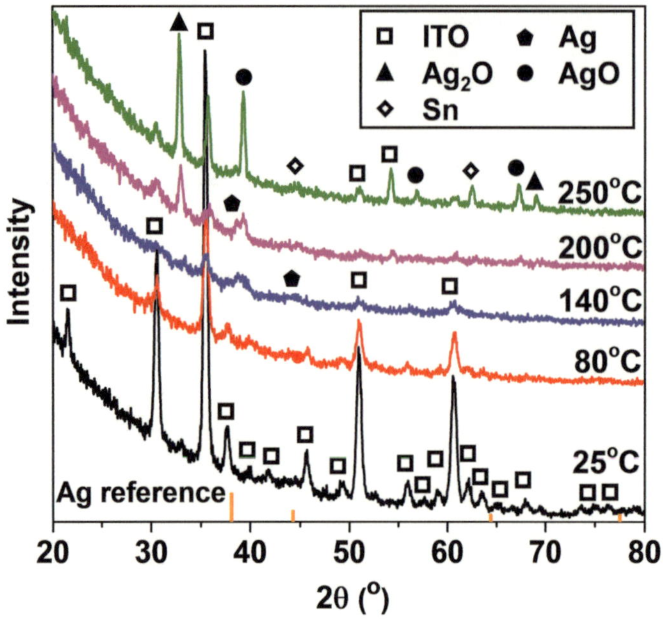

Figure 5.
XRD patterns of Ag nanoparticles deposited on ITO/glass substrates at different substrate temperatures. The reference of bulk Ag is provided in the figure [33].

ejection of oxygen adatoms from the ITO surface by hydrogen etching in ambient plasma at higher substrate temperature leading to oxidation of the Ag nanoparticle surface. Thus, the Ag oxide is formed as a shell and the metallic Ag is as a core of the nanoparticles. The crystallite sizes of Ag_2O are 4.04, 7.94, 13.19, 17.92 and 21.13 nm for different substrate temperatures of 25, 80, 140, 200 and 250°C, respectively, while the crystallite sizes of AgO appeared at 140, 200 and 250°C to be 7.29, 15.29 and 22.45 nm, respectively. The increasing in the crystallite sizes of the Ag oxides is due to the increasing in the diffusion of oxygen adatoms into Ag nanoparticles during the deposition processes. In addition, small diffraction peaks at 44.9 and 62.5° were obviously appeared at 250°C. These diffraction peaks belong to the crystalline planes of metallic Sn (according to JCPDS card no.00-004-0673 for crystalline plane of (211) and (112), respectively). These metallic Sn diffraction peaks indicate to the formation of Sn particles due to the hydrogen plasma etching effect at higher substrate temperature.

The chemical materials of the deposited Ag nanoparticles on ITO surface can be obtained by a wide scan of X-ray photoemission spectroscopy (XPS), as shown in **Figure 6(a)**. The wide scan of XPS of the deposited Ag nanoparticles on ITO substrate exhibited the presence of ITO, Ag and carbon materials. The existence of the carbon material is attributed to adventitious carbon which is normally present in all air-exposed materials. Thus, PAHFE technique is a suggested method to deposit pure Ag nanoparticles without any additional materials which usually exist during the synthesis process. The narrow-scan of Ag materials for different substrate temperatures is presented in **Figure 6(b)**. The highest XPS peak was at 80°C due to the highest density of the nanoparticles. Then, the XPS peaks decrease with the substrate temperature more than 80°C and disappeared at 200 and 250°C. This means that Ag nanoparticles diffuse away from the ITO surface at higher temperatures due to incorporation of SnO and In_2O_3 particles as shell layers onto the Ag nanoparticles. Li et al. have reported decreasing in the Ag content at heat-treatment temperature above 300°C due to the diffusion of Ag nanoparticles away

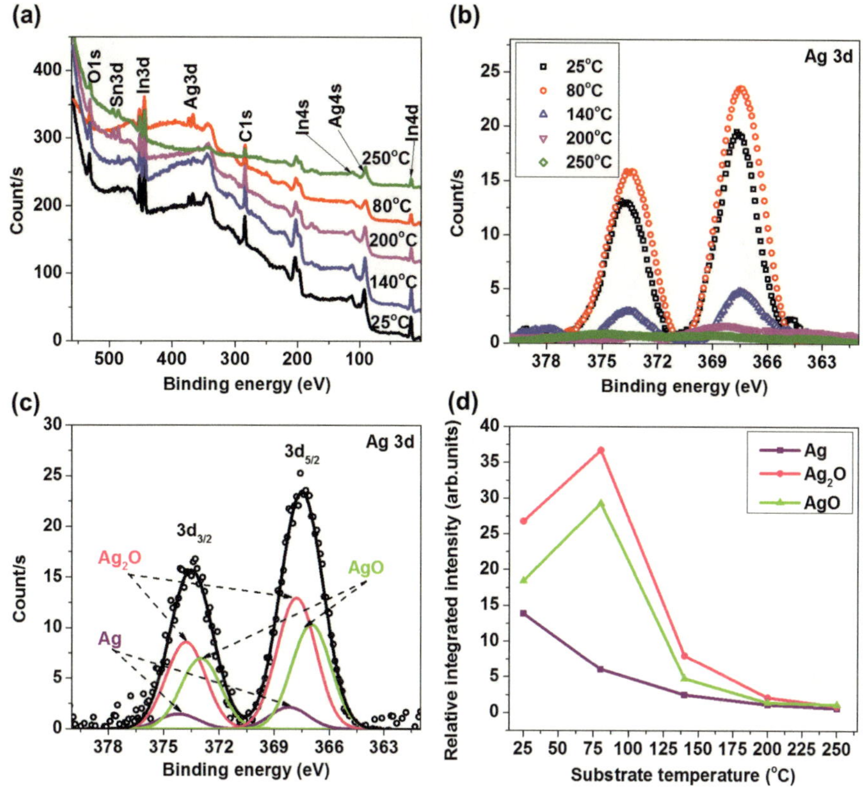

Figure 6.
(a) Wide-scan and (b) Ag 3d narrow-scan of the XPS spectra for Ag nanoparticles deposited on ITO/glass substrates at different substrate temperatures. (c) Typical deconvoluted components to their binding energies of Ag 3d narrow-scan, and (d) variations of the respective deconvoluted components against the substrate temperature.

from surface [42]. In **Figure 6(c)**, the XPS peak of Ag 3d was deconvoluted into three components located at 368.2, 367.8, and 367 eV (Ag $3d_{5/2}$), corresponding to the Ag's metallic state and the two oxidation states of AgO and Ag_2O [43–45]. These Ag oxidation states caused by the formation of AgO and Ag_2O attributed to the diffusion of surface oxide from the ITO substrate, forming a metallic oxide layer on the nanoparticles. The variations of relative integrated intensity for the Ag deconvoluted components are plotted against substrate temperatures in **Figure 6(d)**. The oxidation of Ag nanoparticles is maximal at the substrate temperature of 80°C, forming AgO and Ag_2O shell layers on the nanoparticles and shows decreasing with increasing in the substrate temperature due to the covering by SnO and In_2O_3 materials on the Ag nanoparticles as shell layers at higher temperatures.

3.3 Optical properties

Optical properties of Ag nanoparticles layer are dependent on their sizes and interparticle separations which are dependent on the experimental parameters. The optical absorbance and transmittance can be examined by UV–VIS–NIR spectrophotometer. **Figure 7** showed the optical absorbance and transmittance of the deposited Ag nanoparticles on ITO substrates with using plasma and no plasma conditions at room temperature of the substrate temperature. The absorbance spectra showed one peak for deposited Ag nanoparticles with plasma and three peaks for no plasma condition. This means the deposited Ag nanoparticles at plasma condition appeared as isolated nanoparticles while at no plasma condition

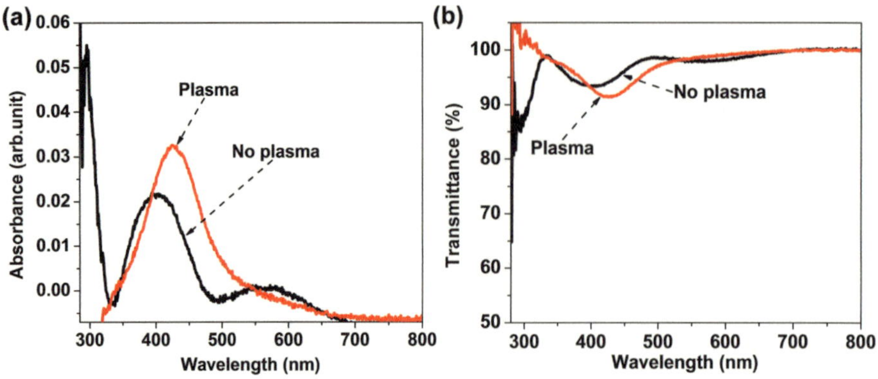

Figure 7.
(a) Absorbance and (b) transmittance spectra of Ag nanoparticles deposited on ITO with plasma and no plasma conditions at room temperature of the substrate temperature.

exhibited a coupled nanoparticles for the two shorter wavelength with appearance larger particles for the third wavelength. The isolated Ag nanoparticles have larger interparticle separations than its diameter while the coupled nanoparticles have smaller interparticle separation than its diameter [46]. The LSPR peak of these isolated Ag nanoparticles deposited with plasma condition showed wide spectrum and centred around 426 nm. However, the nanoparticles size is slightly less than 10 nm that leads to the sharp LSPR peak to be appeared around 400 nm [46]. These shift and wideness are due the formation of Ag oxide as a shell around Ag nanoparticles [8, 47]. Moreover, the intensity of spectrum for with plasma condition appeared higher than no plasma condition that means the deposition higher nanoparticles density at with plasma condition.

For the different substrate temperature with using plasma parameters, the optical absorbance and transmittance spectra of the deposited Ag nanoparticles on ITO substrates are depicted in **Figure 8(a** and **b)**. The SPR peak located at 320 nm remains constant with the change in substrate temperature owing to the absorption coming from the ITO/glass substrate [48]. The SPR peak at 140°C showed slight blue shift with higher intensity compared to at 80°C of the substrate temperature. This slight blue shift is due to a reduction in the interparticle separation to diameter ratio with increasing in the diameter, while the higher intensity belongs to higher area filling fractions for Ag nanoparticles [49]. The absorbance spectra at higher substrate temperature appeared interference fringes in the visible region due to coherent multiple reflections on the ITO surface. The appearance of these reflections comes from the ITO surface decomposition via hydrogen plasma with hot-filament at above 80°C leading to increase in the interference and absorption peaks of ITO with temperature. The transmission spectra show high transmittance at 80 and 140°C and very low at substrate temperature up to 200°C and 250°C. This low transmittance belong to increase in the nanoparticles size leading to larger scattering of light. The interference of ITO decomposition and absorption peaks of Ag nanoparticles were fitted by Gaussian fitting components as demonstrated in **Figure 8(c)**. The appearance of two significant SPR peaks is due to the assembly of Ag nanoparticles as coupled nanoparticles.

The SPR peaks of the deposited Ag nanoparticles on ITO surface in different plasma deposition and Ag wire sizes are shown in **Figure 9(a** and **b)**, respectively. The SPR peak increases with red shift for increasing in the plasma deposition of 12, 18 and 24 W at room temperature of the substrate temperature, as presented in **Figure 9(a)**. The high intensity refers to higher area filling fractions for the

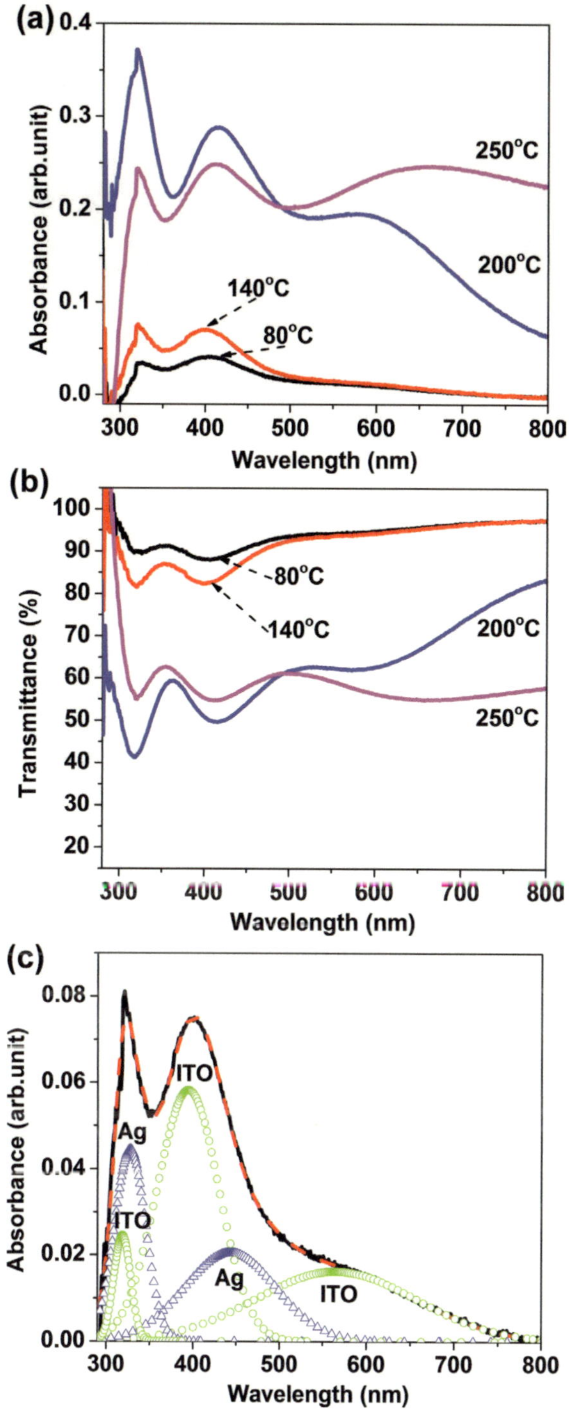

Figure 8.
(a) Absorbance and (b) transmittance spectra of Ag nanoparticles deposited on ITO with using plasma at different substrate temperatures. (c) Gaussian deconvolution of absorbance spectrum for Ag nanoparticles deposited on ITO/glass substrates at 140°C [33].

deposited Ag nanoparticles and the red shift belongs to increases in the nanoparticles size for isolated nanoparticles at 12 and 18 W of plasma deposition. There is a small peak appeared in shorter wavelength at 24 W that means the appearance of

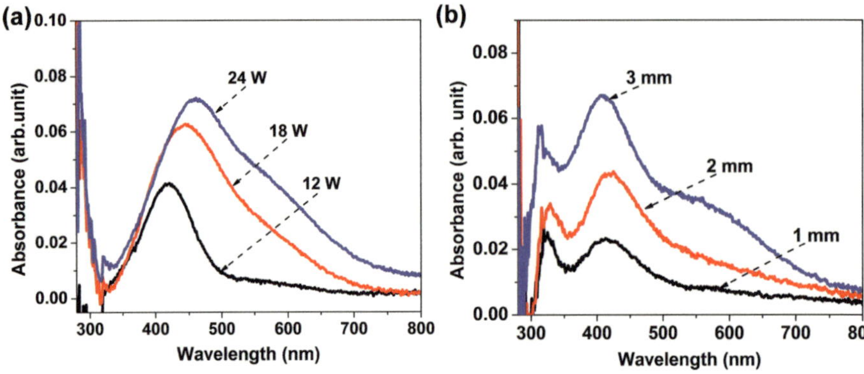

Figure 9.
Absorbance spectra of Ag nanoparticles deposited on ITO surface at (a) different plasma deposition at 25°C, and (b) different Ag wire size at 100°C.

the coupled nanoparticles at higher plasma deposition. For different Ag wire sizes shown in **Figure 9(b)**, the absorbance peaks increase in the intensity with increasing in the Ag wire sizes that belong to increasing in the area filling fractions for the deposited Ag nanoparticles by increasing in the nanoparticles size. The shorter peak showed slight red shift at 2 mm and blue shift at 3 mm of Ag wire size while the middle peak exhibited slight blue shift with appearance of a third peak at 3 mm of Ag wire size. These peaks resulted from the overlapping of the SPR of Ag nanoparticles with the ITO interference due to the substrate temperature of 100°C, and the shifts refer to gradually decreasing in the interparticle separation to diameter ratios at 2 mm and 3 mm, respectively.

3.4 Electronic structure

The optical band gap for ITO with Ag nanoparticles prepared at 25, 80, 140, 200, 250°C of substrate temperatures are about 4.131, 4.097, 4.076, 4.098 and 4.134 eV, respectively [33]. The decreasing of the optical energy gap of the ITO with Ag nanoparticles compared to the determined band gap of the ITO around 4.153 eV is due to the small band gap of Ag_2O and AgO about 1.46 and 1.7 eV, respectively [50, 51]. The small change in the optical band gap is due to the small area filling fraction of the nanoparticles on ITO surface. Moreover, the work function for ITO with Ag nanoparticles prepared at 25, 80, 140, 200, 250°C of substrate temperatures are about 4.18, 3.97, 4.19, 4.29 and 3.26 eV, respectively. These work functions showed slightly lower than the pure ITO work function (4.72 eV) due to the work function of Ag polycrystalline of 4.26 eV [52]. Moreover, the slight reduction in the work function at 80°C is due to the increasing in the crystallite size of Ag_2O, and the slight increment in the work functions at 140 and 200°C are due to the formation of AgO [53]. Furthermore, a reduction in work function to 3.26 eV was noticed in the ITO with Ag nanoparticles prepared at substrate temperature of 250°C. This reduction is due to the decomposition of ITO at higher temperature resulting in the formation of ITO nanoparticles, which have been reported its work function lower than work function of the bulk material (~3.8 eV) [54]. In addition, it is known that the work function is the energy difference between the Fermi level and the vacuum level, and increasing in the work function leads to improvement in electronic conductivity [55]. However, the Ag oxide leads to loss of electrons compared to the metallic Ag due to the reduction in the Fermi level edge [53]. **Figure 10** shows the energy levels diagram of pure ITO and ITO with Ag nanoparticles. The increasing in the energy difference of ITO with Ag nanoparticles compared to pure ITO has

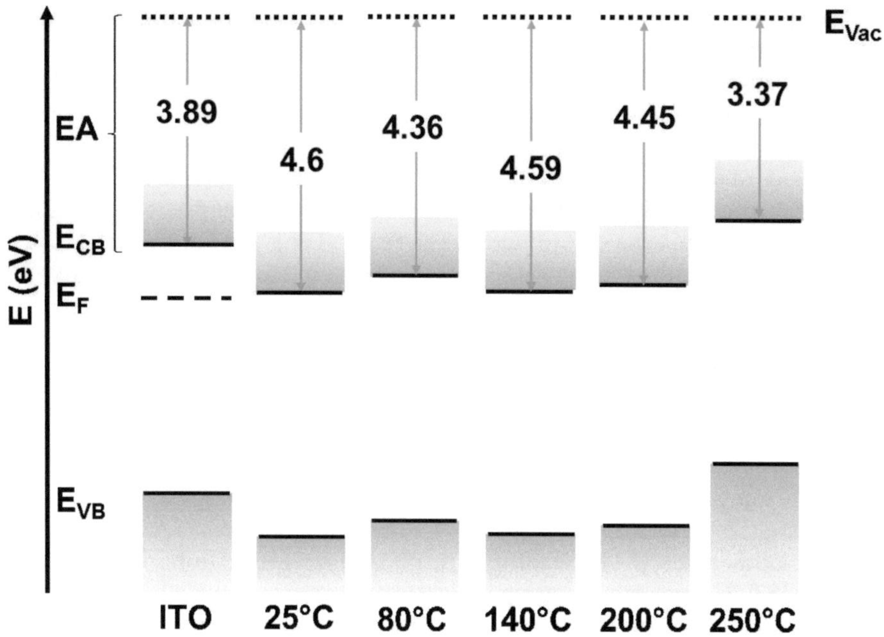

Figure 10.
Energy level diagrams of Ag nanoparticles deposited on ITO/glass substrates at different substrate temperatures and blank ITO substrate, respectively. The valence band energy E_{VB}, Fermi energy E_F, conduction band energy E_{CB}, electron affinity EA, and vacuum level energy E_{Vac} are indicated in the figure [33].

been reported to be null due to an increasing in quantity of free electrons in the Fermi level [33].

3.5 Electrical properties

The distribution of small size of Ag nanoparticles on a layer does not conduct electricity efficiently. However, deposition of these nanoparticles on conducting layer (ITO) may affect on the surface conductivity of the conducting layer. The sheet resistances of ITO surface with Ag nanoparticles prepared with using plasma at different substrate temperatures of 25, 80, 140, 200, 250°C are about 7.75, 7.45, 8, 8.87 and 16.91 Ohm/square, respectively [33]. At the substrate temperature below 200°C, there is no significant effect in the sheet resistances of ITO with Ag nanoparticles and pure ITO (7.68 Ohm/square). However, the sheet resistance of ITO with Ag nanoparticles showed increasing at substrate temperature of 250°C. This increment in the sheet resistance have been reported for ITO particles with thickness of approximately 20 μm to be about 15 ± 5 Ohm/square [56]. Moreover, the resistance sensitivities of AgO and Ag_2O have been reported to be about 50 and 12.5 Ohm.m, respectively [57–59]. Thus, the increasing in the Ag oxidation and the formation of In_2O_3 and SnO particles lead to change in sheet resistance for ITO with Ag nanoparticles at 250°C resulting in increasing in the contact resistance between the Ag nanoparticles and ITO.

4. Growth mechanism

The formation of Ag nanoparticles at different substrate temperatures using PAHFE technique was sequentially clarified in two processes, plasma cleaning and plasma deposition. For the plasma cleaning, a high energetic hydrogen ions as a hydrogen plasma was run for 10 minutes. This hydrogen plasma acts to remove the oxide contaminations from the ITO and Ag wires surfaces as well as activate the substrate surfaces [60, 61]. After the plasma cleaning, high density of remaining surface bonded oxygen was produced on the ITO top and sup-layers and increases with rising in the substrate temperature. Increasing in surface bonded oxygen on the ITO leads to form In_2O_3 and SnO columnar grains at higher temperatures. For plasma deposition, the hot filament was run with the hydrogen plasma for evaporation Ag wire under plasma environment. The thermal energetic of the evaporated Ag adatoms were reached to the ITO surface to form Ag nanoparticles with oxide shell layers of Ag_2O and AgO due to the high-density bonded oxygen surfaces. Moreover, the Ag adatoms have a limited surface mobility on high density bonded oxygen surfaces. Thus, the low substrate temperatures lead to the nanoparticles formation with oxide shell layers due to low bonding energies of Ag-Ag and Ag-O of more than 163 and 213 eV, respectively [62].

The decomposition of ITO surface into SnO and In_2O_3 particles can be observed at substrate temperature more than 200°C due to the low melting points of In and Sn which are 156.6 and 231.9°C, respectively. The hydrogen plasma facilitates the surface decomposition of ITO leading to increase of oxygen adatoms in the nucleation sites [33]. Moreover, increasing in the substrate temperatures leads to increase in surface diffusion of the surface oxygen adatoms into the decomposed sites of ITO resulting in formation of large particles of In_2O_3 and SnO. Furthermore, the bonding energy of SnO (528 eV) is more stable than In_2O_3 (360 eV) at higher substrate temperature of 250°C. Thus, the oxygen surface diffusion with decomposition of ITO enhances the formation of SnO nanoparticles leading to reducing in the AgO, Ag_2O, and In_2O_3 nanoparticles on the surface at higher substrate temperatures.

5. Conclusion

Ag nanoparticles layer were deposited on ITO substrates via PAHFE technique at low substrate temperatures less than 200°C. This technique is simple and fast deposition which required approximately 3 minutes of the deposition time and showed directly formation of the nanoparticles layer from the pure Ag wire source. The size and interparticle separation of Ag nanoparticles can be controlled by variation in the experimental parameters. The morphological, structural, optical, and electrical properties of the Ag nanoparticles produced were also studied. Using plasma during the deposition leads to formation of uniform size and distribution of Ag nanoparticles. The nanoparticles growth rate was induced by increasing in the substrate temperature up to 140°C leading to increasing in the nanoparticles size and interparticle separation. The oxygen adatoms can diffuse to the nanoparticles to form Ag oxide shell which was enhanced by the plasma ambient and substrate temperature. Moreover, the SPR of the nanoparticles was strongly dependent on the diameter and interparticle separation of the nanoparticles. In addition, the high population of the Ag nanoparticles at 80°C improved the surface conductivity for the ITO with Ag nanoparticles to be at 7.45 Ohm/square and decreased the work function to 3.97 eV. However, the formation of SnO and In_2O_3 particles at the substrate temperature of 250°C increased the sheet resistance to 16.91 Ohm/square and more decreasing in the work function to 3.26 eV. Furthermore, the variation in the experimental parameters using PAHFE technique was observed to tune the optical and electrical properties of ITO with Ag core and Ag oxide shell nanoparticles resulting in changes in the functionality of Ag nanoparticle-coated transparent conductive oxide thin films to be useful in the applications of the optoelectronic devices.

Acknowledgements

This work was supported by the University of Malaya UMRG Programme - Frontier Science (RP038C-17AFR) and the University of Malaya Research Partnership Grant (RK006-2019). We would also like to thank the SLRI BL3.2a beamline staffs that help us in measuring XPS and UPS.

Author details

Abtisam Hasan Hamood Al-Masoodi[1*], Boon Tong Goh[1], Ahmed H.H. Al-Masoodi[2] and Wan Haliza Binti Abd Majid[1]

1 Low Dimensional Materials Research Center, Department of Physics, Faculty of Science, University of Malaya, 50603 Kuala Lumpur, Malaysia

2 Department of Electronic and Telecommunication Engineering, College of Engineering, The American University of Kurdistan, 42001 Duhok, Kurdistan Region of Iraq

*Address all correspondence to: abtalmasoodi@gmail.com

References

[1] Ko S-J, Choi H, Lee W, Kim T, Lee BR, Jung J-W, et al. Highly efficient plasmonic organic optoelectronic devices based on a conducting polymer electrode incorporated with silver nanoparticles. Energy & Environmental Science. 2013;**6**:1949-1955

[2] Liang Z, Sun J, Jiang Y, Jiang L, Chen X. Plasmonic enhanced optoelectronic devices. Plasmonics. 2014;**9**:859-866

[3] Naghib SM. Fabrication of nafion/silver nanoparticles/reduced graphene nanosheets/glucose oxidase nanobiocomposite for electrochemical glucose biosensing. Anal. Bioanal. Electrochem. 2016;**8**:453-465

[4] Awazu K, Fujimaki M, Rockstuhl C, Tominaga J, Murakami H, Ohki Y, et al. A plasmonic photocatalyst consisting of silver nanoparticles embedded in titanium dioxide. Journal of the American Chemical Society. 2008;**130**:1676-1680

[5] Fukuura T. Plasmons excited in a large dense silver nanoparticle layer enhance the luminescence intensity of organic light emitting diodes. Applied Surface Science. 2015;**346**:451-457

[6] Kalfagiannis N, Karagiannidis P, Pitsalidis C, Panagiotopoulos N, Gravalidis C, Kassavetis S, et al. Plasmonic silver nanoparticles for improved organic solar cells. Solar Energy Materials and Solar Cells. 2012;**104**:165-174

[7] Noguez C. Surface plasmons on metal nanoparticles: The influence of shape and physical environment. The Journal of Physical Chemistry C. 2007;**111**:3806-3819

[8] Schinca D, Scaffardi L, Videla F, Torchia G, Moreno P, Roso L. Silver–silver oxide core–shell nanoparticles by femtosecond laser ablation: Core and shell sizing by extinction spectroscopy. Journal of Physics D: Applied Physics. 2009;**42**:215102

[9] Lok C-N, Ho C-M, Chen R, He Q-Y, Yu W-Y, Sun H, et al. Silver nanoparticles: Partial oxidation and antibacterial activities. JBIC Journal of Biological Inorganic Chemistry. 2007;**12**:527-534

[10] Lee J, Jung B-J, Lee J-I, Chu HY, Do L-M, Shim H-K. Modification of an ITO anode with a hole-transporting SAM for improved OLED device characteristics. Journal of Materials Chemistry. 2002;**12**:3494-3498

[11] Kuwabara T, Nakayama T, Uozumi K, Yamaguchi T, Takahashi K. Highly durable inverted-type organic solar cell using amorphous titanium oxide as electron collection electrode inserted between ITO and organic layer. Solar Energy Materials and Solar Cells. 2008;**92**:1476-1482

[12] Kimoto K, Kamiya Y, Nonoyama M, Uyeda R. An electron microscope study on fine metal particles prepared by evaporation in argon gas at low pressure. Japanese Journal of Applied Physics. 1963;**2**:702

[13] Park ST, Kim T-H, Park D-W. Influence of injected silver content on synthesis of silver coated nickel particles by DC thermal plasma. Applied Surface Science. 2016;**374**:257-264

[14] Bogle K, Dhole S, Bhoraskar V. Silver nanoparticles: Synthesis and size control by electron irradiation. Nanotechnology. 2006;**17**:3204

[15] Suh J, Han B, Kim DS, Choi M. A method for enhanced charging of nanoparticles via condensation

magnification. Journal of Aerosol Science. 2005;**36**:1183-1193

[16] Azizian-Kalandaragh Y, Nouhi S, Amiri M. Effect of post-annealing treatment on the wetting, optical and structural properties of Ag/indium tin oxide thin films prepared by electron beam evaporation technique. Materials Express. 2015;**5**:137-145

[17] Nazarudin NFFB, Azizan SNAB, Rahman SA, Goh BT. Growth and structural property studies on NiSi/SiC core-shell nanowires by hot-wire chemical vapor deposition. Thin Solid Films. 2014;**570**:243-248

[18] Al-Masoodi AHH, Hamzan NB, Al-Masoodi AHH, Rahman SA, Goh BT. Influences of hydrogen dilution on the growth of Si-based core–shell nanowires by HWCVD, and their structure and optical properties. Applied Physics A. 2016;**122**:239

[19] Teo KB, Hash DB, Lacerda RG, Rupesinghe NL, Bell MS, Dalal SH, et al. The significance of plasma heating in carbon nanotube and nanofiber growth. Nano Letters. 2004;**4**:921-926

[20] Thomas R, Mantini M, Rudder R, Malta D, Hattangady S, Markunas R. Carbon and oxygen removal from silicon (100) surfaces by remote plasma cleaning techniques. Journal of Vacuum Science & Technology A: Vacuum, Surfaces, and Films. 1992;**10**:817-822

[21] Baker M. Plasma cleaning and the removal of carbon from metal surfaces. Thin Solid Films. 1980;**69**:359-368

[22] Mathur S, Sivakov V, Shen H, Barth S, Cavelius C, Nilsson A, et al. Nanostructured films of iron, tin and titanium oxides by chemical vapor deposition. Thin Solid Films. 2006;**502**:88-93

[23] Nazarudin NFFB, Noor NJBM, Rahman SA, Goh BT. Photoluminescence and structural properties of Si/SiC core–shell nanowires growth by HWCVD. Journal of Luminescence. 2015;**157**:149-157

[24] binti Hamzan N, bin Ramly MM, Huang NM, Rahman SA, Goh BT. Growth of high density NiSi/SiC core-shell nanowires by hot-wire chemical vapour deposition for electrochemical applications. Materials Characterization. 2017;**132**:187-198

[25] Goh BT, Rahman SA. Study of the growth, and effects of filament to substrate distance on the structural and optical properties of Si/SiC core–shell nanowires synthesized by hot-wire chemical vapor deposition. Materials Chemistry and Physics. 2014;**147**:974-981

[26] Tehrani FS, Goh BT, Muhamad MR, Rahman SA. Pressure dependent structural and optical properties of silicon carbide thin films deposited by hot wire chemical vapor deposition from pure silane and methane gases. Journal of Materials Science: Materials in Electronics. 2013;**24**:1361-1368

[27] Alizadeh M, Ooi PC, bin Omar MF, Dee CF, Goh BT. Solid-state limited nucleation of NiSi/SiC core-shell nanowires by hot-wire chemical vapor deposition. Materials. 2019;**12**:674

[28] Li D, Bulou S, Gautier N, Elisabeth S, Goullet A, Richard-Plouet M, et al. Nanostructure and photocatalytic properties of TiO2 films deposited at low temperature by pulsed PECVD. Applied Surface Science. 2019;**466**:63-69

[29] Goh BT, Wah CK, Aspanut Z, Rahman SA. Structural and optical properties of nc-Si: H thin films deposited by layer-by-layer technique. Journal of Materials Science: Materials in Electronics. 2014;**25**:286-296

[30] Goh BT, Ngoi SK, Yap SL, San Wong C, Dee CF, Rahman SA. Structural and optical properties of the nc-Si: H thin films irradiated by high energetic ion beams. Journal of Non-Crystalline Solids. 2013;**363**:13-19

[31] Rahman MAA, Goh BT, Chiu WS, Haw CY, Mahmood MR, Khiew PS, et al. Aging-and thermal-annealing effects on the vibrational-and microstructural-properties of PECVD grown hydrogenated amorphous silicon carbon nitride thin films. Vibrational Spectroscopy. 2018;**94**:22-30

[32] Hu Z, Li Z, Xia Z, Jiang T, Wang G, Sun J, et al. PECVD-derived graphene nanowall/lithium composite anodes towards highly stable lithium metal batteries. Energy Storage Materials. 2019;**22**:29-39

[33] Al-Masoodi AHH, Nazarudin NFFB, Nakajima H, Tunmee S, Goh BT, Majid WHBA. Controlled growth of silver nanoparticles on indium tin oxide substrates by plasma-assisted hot-filament evaporation: Physical properties, composition, and electronic structure. Thin Solid Films. 2020;**693**:137686

[34] Chang G, Zhang J, Oyama M, Hirao K. Silver-nanoparticle-attached indium tin oxide surfaces fabricated by a seed-mediated growth approach. The Journal of Physical Chemistry B. 2005;**109**:1204-1209

[35] Kuo YL, Juang TY, Chang SH, Tsai CM, Lai YS, Yang LC, et al. Influence of temperature on the formation of silver nanoparticles by using a seed-free photochemical method under sodium-lamp irradiation. ChemPhysChem. 2015;**16**:3254-3263

[36] Dang MC, Dang TMD, Fribourg-Blanc E. Silver nanoparticles ink synthesis for conductive patterns fabrication using inkjet printing technology. Advances in Natural Sciences: Nanoscience and Nanotechnology. 2014;**6**:015003

[37] Xiliang Q, Yang C, Tiesong L, Peng H, Jun W, Ping L, et al. Large-scale synthesis of silver nanoparticles by aqueous reduction for low-temperature sintering bonding. Journal of Nanomaterials. 2014;**2014**:8

[38] Jiang X, Chen W, Chen C, Xiong S, Yu A. Role of temperature in the growth of silver nanoparticles through a synergetic reduction approach. Nanoscale Research Letters. 2011;**6**:32

[39] Zhou D, Bennett SW, Keller AA. Increased mobility of metal oxide nanoparticles due to photo and thermal induced Disagglomeration. PLoS One. 2012;7:e37363

[40] Moon J-M, Bae J-H, Jeong J-A, Jeong S-W, Park N-J, Kim H-K, et al. Enhancement of hole injection using ozone treated Ag nanodots dispersed on indium tin oxide anode for organic light emitting diodes. Applied Physics Letters. 2007;**90**:163516

[41] Wei H, Eilers H. From silver nanoparticles to thin films: Evolution of microstructure and electrical conduction on glass substrates. Journal of Physics and Chemistry of Solids. 2009;**70**:459-465

[42] Li W, Seal S, Megan E, Ramsdell J, Scammon K, Lelong G, et al. Physical and optical properties of sol-gel nano-silver doped silica film on glass substrate as a function of heat-treatment temperature. Journal of Applied Physics. 2003;**93**:9553-9561

[43] Kim SH, Choi WI, Kim KH, Yang DJ, Heo S, Yun D-J. Nanoscale chemical and electrical stabilities of graphene-covered silver nanowire networks for transparent conducting

electrodes. Scientific Reports. 2016;**6**:33074

[44] Milczarek G, Rebis T, Fabianska J. One-step synthesis of lignosulfonate-stabilized silver nanoparticles. Colloids and Surfaces B: Biointerfaces. 2013;**105**:335-341

[45] Hoflund GB, Hazos ZF, Salaita GN. Surface characterization study of Ag, AgO, and Ag_2O using x-ray photoelectron spectroscopy and electron energy-loss spectroscopy. Physical Review B. 2000;**62**:11126

[46] Agnihotri S, Mukherji S, Mukherji S. Size-controlled silver nanoparticles synthesized over the range 5-100 nm using the same protocol and their antibacterial efficacy. RSC Advances. 2014;**4**:3974-3983

[47] Schinca D, Scaffardi L. Core and shell sizing of small silver-coated nanospheres by optical extinction spectroscopy. Nanotechnology. 2008;**19**:495712

[48] Yun S, Hong S, Acapulco JA, Jang HY, Ham S, Lee K, et al. Close-packed two-dimensional silver nanoparticle arrays: Quadrupolar and dipolar surface plasmon resonance coupling. Chemistry-A European Journal. 2015;**21**:6165-6172

[49] Kinnan MK, Chumanov G. Plasmon coupling in two-dimensional arrays of silver nanoparticles: II. Effect of the particle size and interparticle distance. The Journal of Physical Chemistry C. 2010;**114**:7496-7501

[50] Kumar GA, Reddy MR, Reddy KN. Structural and optical properties of AgO thin films grown by RF reactive magnetron sputtering technique, advanced nanomaterials and emerging engineering technologies (ICANMEET). In: 2013 International Conference on, IEEE. 2013. pp. 354-356

[51] Ida Y, Watase S, Shinagawa T, Watanabe M, Chigane M, Inaba M, et al. Direct electrodeposition of 1.46 eV bandgap silver (I) oxide semiconductor films by electrogenerated acid. Chemistry of Materials. 2008;**20**:1254-1256

[52] Chen S-H, Li Y-R, Yu C-F, Lin C-F, Kao P-C. Enhanced luminescence efficiency of Ag nanoparticles dispersed on indium tin oxide for polymer light-emitting diodes. Optics Express. 2013;**21**:26236-26243

[53] Akbi M. Effect of arcing in air on the photoelectric work function of silver-based contacts. IEEE Transactions on Plasma Science. 2015;**43**:637-642

[54] Kim C, Kim YH, Noh YY, Hong SJ, Lee MJ. Improved charge injection of metal oxide thin-film transistors by stacked electrodes of indium tin oxide nanoparticles and silver nanowires. Advanced Electronic Materials. 2018;**4**:1700440

[55] Klein A, Körber C, Wachau A, Säuberlich F, Gassenbauer Y, Harvey SP, et al. Transparent conducting oxides for photovoltaics: Manipulation of fermi level, work function and energy band alignment. Materials. 2010;**3**:4892-4914

[56] Bühler G, Thölmann D, Feldmann C. One-pot synthesis of highly conductive indium tin oxide nanocrystals. Advanced Materials. 2007;**19**:2224-2227

[57] Nwanya AC, Ugwuoke PE, Ezekoye BA, Osuji RU, Ezema FI. Structural and optical properties of chemical bath deposited silver oxide thin films: Role of deposition time. Advances in Materials Science and Engineering. 2013;**2013**:8

[58] Varkey A, Fort A. Some optical properties of silver peroxide (AgO) and silver oxide (Ag2O) films produced

by chemical-bath deposition. Solar Energy Materials and Solar Cells. 1993;**29**:253-259

[59] Jeong J-A, Lee J, Kim H, Kim H-K, Na S-I. Ink-jet printed transparent electrode using nano-size indium tin oxide particles for organic photovoltaics. Solar Energy Materials and Solar Cells. 2010;**94**:1840-1844

[60] Chong SK, Goh BT, Aspanut Z, Muhamad MR, Dee CF, Rahman SA. Synthesis of indium-catalyzed Si nanowires by hot-wire chemical vapor deposition. Materials Letters. 2011;**65**:2452-2454

[61] Chong SK, Goh BT, Aspanut Z, Muhamad MR, Dee CF, Rahman SA. Radial growth of slanting-columnar nanocrystalline Si on Si nanowires. Chemical Physics Letters. 2011;**515**:68-71

[62] Maruyama T, Arai S. Electrochromic properties of cobalt oxide thin films prepared by chemical vapor deposition. Journal of the Electrochemical Society. 1996;**143**:1383-1386

Section 4

Photovoltaic Thin Films

Chapter 8

Solution-Processed Chalcogenide Photovoltaic Thin Films

Marcos Antonio Santana Andrade Junior,
Hugo Leandro Sousa dos Santos, Mileny dos Santos Araujo,
Arthur Corrado Salomão and Lucia Helena Mascaro

Abstract

Chalcogenides-based thin film solar cells are great competitors to beat high efficiencies as silicone solar cells. The chalcogenides that have been commonly used as absorber materials are CIS, CIGS, and CZTS. They present some advantages of having a direct and tunable band gap, high absorption coefficient and respectable efficiency to cost ratio. Solution processable deposition approaches for the fabrication of solar cells attracts a great deal attention due to its lower capital cost of the manufacturing than the vacuum-based techniques. In this chapter, we detail the use of a low-cost method of deposition for the chalcogenide thin films by spin-coating and spray-coating, which is already widely employed in several fields of industries.

Keywords: chalcogenides, solar cells, spin-coating, spray

1. Introduction

Currently, the photovoltaic market is based on silicon solar cells with conversion efficiency of 15–22%. However, chalcogenides-based thin films solar cells are great competitors of silicon technologies. Among of all photovoltaic thin films, chalcogenide solar cells present some advantages of having a direct and tunable band gap (1.0–2.5 eV), high absorption coefficient ($>10^4$ cm^{-1}) and respectable efficiency to cost ratio [1]. Devices containing $CuInGa(S,Se)_2$ (CIGS) recently recorded a 23% efficiency [2] and it becomes possible to PV modules based on that chalcogenide thin film to have already entered the market at similar or even lower costs than traditional silicon modules [3]. Despite this high efficiency level, the CIGS-based PV technology has not yet attained its full potential. If all loss mechanism were addressed at the same time, the theoretical maximum 30% could be technically feasible. The reported highest efficiency for CIGS thin film solar cell is based on vacuum process for coevaporating the elements and depositing the absorber layer. However, vacuum-based methods are quite expensive, which creates cost and technological barriers to produce low-cost photovoltaic modules [4]. Therefore, some approaches have been investigated for further improvements, as well as, to develop cheaper strategies for the absorber layer.

Solution-processed techniques have been extensively studied to deposit chalcogenide thin films applied to the second-generation solar cells. The characteristics of the precursor solutions are fundamental to perform the deposition and have an important role in the resulting film. The solution composition, concentration of constituents, viscosity, and solvent will influence on the film adherence onto substrate, grain growth, and most importantly, on the solar cell efficiency.

The solution processable deposition of chalcogenides absorber layer, such as CIS ($CuIn(S,Se)_2$), CZTS ($Cu(Zn,Sn)(S,Se)_2$), and CIGS compounds for the fabrication of solar cells attracts a great deal attention due to its lower capital cost of the manufacturing than the vacuum-based approaches, high production rate, ability for roll-to-roll production, compositional uniformity over large areas, and high material utilization [5, 6]. Several solution processed methods are already commonly used for chalcogenides film deposition, such as, electrodeposition [4], spin coating [7], ink jet printing, electrophoretic deposition [8], etc.

Among the solvents used to prepare solution-processed high-performance solar cells reported in literature are water (H_2O), dimethyl sulfoxide (C_2H_6OS), hydrazine (N_2H_4), methanol (CH_3OH), dimethylformamide (C_3H_7NO) [9]. The solvent used in the precursor solution must be environmental-friendly, being able to dissolve the salts used as source of the chalcogenide cations and the compounds that are the source of the chalcogens, and not contribute to impurities in the film [10].

This chapter discusses the two most reported techniques used to prepare the world-record efficiencies of solution-processed chalcogenide solar cells: spin-coating and spray-coating. The fundamental characteristics of the solutions, regarding the physical chemical properties of the solvents, and the important characteristics of the deposition methods will be discussed.

2. Absorber layers deposited by spin-coating

The spin coating is a technique to deposit thin films on flat substrates. It has been very utilized on the fabrication of films in thickness range of micrometer to nanometer. This makes it attractive to prepare solar cells, on a manufactory point of view, mainly at laboratorial scale.

The process to produce thin films with this method can be resumed basically in three steps (**Figure 1** represents a schematic diagram of the steps), respectively; fluid dispense on the substrate, spreading of fluid (spin up followed by spin off) and evaporation. The solution dispensed spreads over the substrate surface by the centrifugal force and, at last, the evaporation forms the film.

The viscosity of the fluid dispensed (η), density of volatile liquid (ρ_A), evaporation rate (m), and angular speed of the spinning plate (ω) are the main factors that

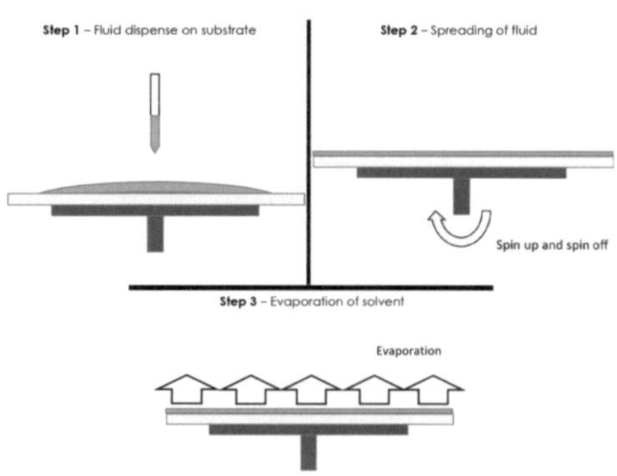

Figure 1.
Steps of spin coating technique.

affects the layer thickness, the following equation represents the variation of the thickness in function of this parameters [11]:

$$h = \left(1 - \frac{\rho_A}{\rho_{A0}}\right) \cdot \left(\frac{3\eta . m}{2\rho_{A0}\omega^2}\right)^{1/3} \quad (1)$$

where thickness is h. These factors and possible contaminants (oxygen, humidity, solvent traces, etc.) can also influence on the roughness and the uniformity of the film [12].

Spin coating is a fast and low-cost process. The feasibility of the process is due to cheap material required, comparing to vacuum-based coatings, since a spinning plate is much cheaper than vacuum system [12]. In addition, this method allows the use of a wide range of particle diameters, which means that the method can be utilized for different applications. On the solar cells, smaller particles diameters mean more grain boundaries, and consequently, loss of charge carrier [13].

Otherwise, it is important to note that at the spin coating process, there is an expressive material wastage, since only 2–5% of the material stays on the layers, while 95–98% drops out of the layer [14]. Another disadvantage is that this technique does not form uniform layers over a large area and it does not have roll-to-roll capability [15]. Lastly, there is a difficulty at the production of multilayers devices with this method.

The spin coating has a good cost effective, mainly at laboratorial use, to produce thin and uniform films. Anyway, it is indispensable to consider the advantages and the disadvantages before using this method to conclude if this is viable.

Since the success of the deposition of $Sn(S,Se)_x$ films using hydrazine-based solution, this method has been applied to fabricate highly efficient devices. CIGS thin film was deposited by spin coating using a hydrazine-based solution containing the chloride salts of the cations and thiourea [16]. Hydrazine is a solvent that has been used to prepare molecular inks to deposit highly pure films for highly efficient solar cells. The current world record efficiency for a solution-processed solar cell is hold by a device containing CIGS deposited via hydrazine-based solution (12.6% for a CIGS solar cell) [17, 18]. Hydrazine is used to dissolve elemental metals and binary compounds, as SnSe to produce kesterite solar cells [19]. It has also been used to prepare precursor solutions containing Cu_2S e In_2Se_3 to deposit $CuIn(S,Se)_2$ thin films [20]. The advantage of using hydrazine is because it is a poor coordination solvent. During thermal treatment, hydrazine decomposes in small molecules (H_2, NH_3 and N_2) which easily leaves the film without keeping residues. The absence of carbon or oxygen in hydrazine structure avoid organic impurities which compromise the solar cell efficiency [21]. However, hydrazine use is limited because it is highly toxic irritant, highly reactive and easily catch fire [12]. Because of that, hydrazine-free solutions have been studied to unexpected dangerous reactions.

Alcohols are a green, less toxic, and low-cost alternative for deposition solutions. Methanol-based solution containing the chloride salts of the cations and thiourea, as sulfur source, have been used to deposit CIGS films. The cation acetate salts are more recommended rather than the nitrate and chloride-based salts to form absorber layers with a minimum of residues.

Alcoholic solutions have also been used to prepare inks with butyldithiocarbamic acid (BDCA) as the sulfur source instead of thiourea to deposit CZTS solar cells with 6% conversion efficiency. Ethanol-based solutions with BCDA to dissolve binary oxides Cu_2O, ZnO and SnO to deposit CZTS films [20]. Ammonium thioglycolate has also been used to dissolve metallic oxides as an environmentally friendly alternative [10].

The cations and thiourea complexes with dimethyl sulfoxide (DMSO) or dimethyl formamide (DMF) to prepare molecular inks to deposit chalcogenide absorber layers [22]. This combination acts stabilizing the desired oxidation state of the metals in the resulting film [10]. Thiourea binds to the cations and avoid their evaporation during annealing [23, 24]. DMF and DMSO also dissolves selenourea to prepare the selenide compounds [9].

Beyond the salt-based precursor solutions and molecular inks, nanocrystals inks have also been used to deposit chalcogenide absorber films. CZTS solar cells fabricated using a nanoparticle-based solution have presented a maximum of 10.2% conversion efficiency, which is close value to the efficiency for a device prepared using hydrazine. The synthesis of the nanocrystals typically consist of oleylamine solution of the cation salts (chlorides or acetylacetonate) [13, 25], and a hot injection of sulfur oleylamine solution. Dodecanethiol and hexanethiol are alternatives to oleylamine [20]. Dichlorobenzene has also been used to dissolve sulfur. Dichlorobenzene is not recommended to prepare Se-based chalcogenides because of the Se low solubility in this solvent [25]. After separation and purification, the nanocrystals are dissolved in an organic solvent such ethanol, hexane, chloroform, or toluene to prepare the nanocrystal inks.

The combination of molecular and nanocrystal inks is known as hybrid ink [20]. A hybrid solution prepared by dispersing Cu_2SnS_3 nanocrystals in a solution containing Zn and propylmercaptan to obtain a hybrid solution applied to deposit Cu_2ZnSnS_4 thin films. The hybrid solution-based CZTS solar cell presented a PCE of 5% and Voc of 440 mV [20]. Solvents may leave residues trapped in the film. The time and temperature for annealing and sulfur/selenization are not only essential for crystal growth but must also remove all the possible residues left by solvent and other organic [5, 26].

Although the CIGS devices prepared by vacuum-based techniques present values of efficiencies of ~23%, the efficiency of a spin-coating deposited $CuIn(S,Se)_2$ solar cell is similar to the efficiency of a $Cu(Ga,In)(S,Se)_2$ device fabricated in the same condition. In addition, the efficiency of a CZTS solar cell is the same as for a device fabricated by vacuum techniques. This evidences that spin-coating is an inexpensive alternative to fabricate highly efficient devices. The efforts for developing inks using environmentally friendly solvents, metals and chalcogen precursors, decreasing the residues in the films are fundamental to approximate the photovoltaic characteristics of a solution-processed device to the characteristics of the expensive vacuum-based solar cells (**Table 1**).

Absorber	solvent	V_{OC} (V)	J_{SC} *(mA cm^{-2})*	FF (%)	PCE (%)	Ref.
$Cu(In,Ga)(S,Se)_2$	hydrazine	0.623	32.6	75	15.2	[16]
$Cu(In,Ga)S_2$	methanol	0.787	17.0	62	8.3	[27]
$CuIn(S,Se)_2$	ethanol	0.561	27.6	65	10.1	[28]
CuIn(S,Se)2	DMF	0.512	36.9	71	13.4	[9]
$Cu_2ZnSn(S,Se)_2$	DMSO	0.457	36.0	65	10.7	[29]
$Cu_2Zn(Sn,Ge)(S,Se)_2$	DMF	0.583	33.6	56	11.0	[9]

Table 1.
Photovoltaic parameters of highly efficient chalcogenide solar cells fabricated by spin-coating.

3. Spray-coating: a low-cost technique for chalcogenide solar cells

In general, the success of the deposition of a thin film by spray-pyrolysis depends mainly on three factors: the composition of the precursor solution, the atomization and droplet transport process and the substrate temperature [30]. The control or modification of these parameters directly impacts the quality of the obtained film. Considering the production of the precursor solution, it can be composed of inorganic salts or organometallic compounds dissolved in aqueous or organic solvents, respectively.

The choice of the precursor reagent and solvent is an important step, since it will define the maximum concentration of salt that can be used and will directly affect the process of formation and transport of the aerosol in the atomization step [30]. The use of aqueous solvents and inorganic salts are the most used reagents, since they have lower degree of toxicity and environmental pollution than organic solvents and organometallic compounds. In addition, it has been observed that the use of alcoholic organic solvents can lead to carbonaceous impurities that are harmful to the photovoltaic device. They affect the growth of crystals in the film and acting as barriers for extracting loads, leading to low values of open-circuit voltage (V_{oc}) and fill factor (FF) of the solar cell [31, 32]. A common approach in the production of precursor solutions for deposition of copper-based chalcogenides thin films, such as CIS and CIGS, is to prepare aqueous solutions stocking the chloride or nitrate salts of the metals of interest, and the sulfur source, which is normally the thiourea. The concentration of the sulfur precursor is usually 5 to 10 times more concentrated than metallic salts. The excess of the sulfur has a compensatory role due to losses during deposition, and when using molybdenum substrate, it prevents the formation of oxides of this metal [33, 34]. These solutions are diluted and mixed to obtain the final precursor solution, the concentration of each component in the final solution must be varied in order to assess the influence of each metal on the chemical and physical properties of the obtained films. Solar cells with a CIGS absorbing layer obtained by spray-pyrolysis using this approach in the preparation of the precursor solution have achieved 5 to 10% efficiency [31, 35].

The typical atomizers used for depositing chalcogenides films are the pneumatic and ultrasonic. The aerosol formation mechanism is quite different between the two models, while in the pneumatic the aerosol is formed by the action of a pressurized gas, in the ultrasonic, the aerosol is produced by ultrasonic waves generated by a piezoelectric component that is in contact with the solution, the formed aerosol is transported to the substrate surface by a carrier gas [36, 37]. Although ultrasonic atomizers are more expensive and complex than pneumatic ones, these atomizers have better control of average droplet size and the rate of aerosol formation is independent of the flow rate of the carrier gas. The efficient control of the droplet size distribution is important due to its influence on uniformity of the deposited film [38]. The two types of atomizers mentioned are also differentiated by the atomization rate and the initial speed of the drops, because as the pneumatic works with a pressurized gas carrier, both the atomization rate and the velocity of the droplets are much higher [30, 39]. In pneumatic spraying, the main factor to be tested is the pressure of the carrier gas, since the distribution of the average size of the droplets and the initial velocity of the droplets, as well as the rate of aerosol formation are directly linked to this factor. On the other hand, when using an ultrasonic atomizer the main factor to be monitored is the atomization frequency, since the distribution of the average droplet size is defined by this parameter [40]. When choosing carrier gas, one must consider whether it is reactive or not. Oxidizing gases, such as O_2, should be avoided, as they can react with the precursor metals in the pyrolysis phase,

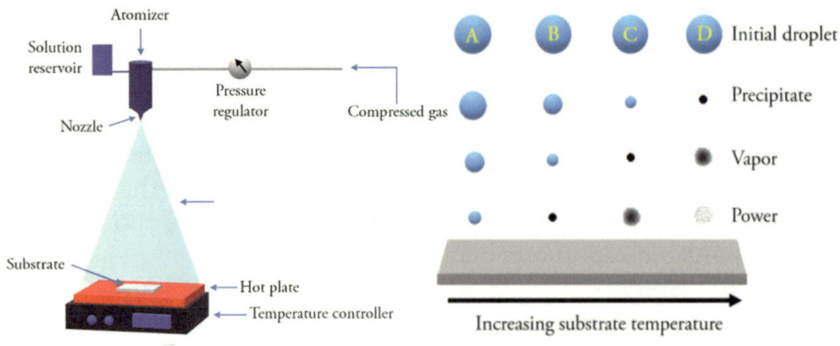

Figure 2.
Schematic illustration of thin film deposition by spray-pyrolysis and the effect of increasing substrate temperature on the spray-pyrolysis deposition.

leading to the formation of unwanted oxides. In this sense, inert gases such as N_2 or Ar are the most suitable for depositing chalcogenides by spray-pyrolysis [41].

Among the several factors that must be taken when depositing thin films by spray-pyrolysis, the substrate temperature is pointed out as the most important parameter [38]. Many stages of the thin film deposition process depend directly on the substrate temperature, among which are: droplet spreading, solvent evaporation and salt decomposition and compound formation. All these steps are important for obtaining the compound with the desired chemical and physical characteristics in the form of a homogeneous thin film. **Figure 2** summarizes the effects of increasing the temperature on the deposition process [30, 38, 39]. At low temperatures, the solvent still does not evaporate and the droplets of the precursor solution collide with the substrate and undergo decomposition (process A). At intermediate temperatures the solvent evaporates completely during transport, reaching the substrate in the form of a dry precipitate which then decomposes (process B). From intermediate to high temperatures, the solvent evaporates, and the solid precipitate is formed, but before reaching the substrate the precipitate melts and vaporizes, leading to chemical vapor deposition (CVD) (process C). Finally, at high temperatures (process D) the vapor phase formed after melting the precipitate decomposes to form the compound powder, which is deposited on the substrate. In this context, the processes A and D are not indicated for chalcogenides deposition, because the films obtained are often rough or the form of poorly adherent powder [38]. Thus, temperatures where processes B or C can occur (300° to 450°C), are the most suitable for the deposition of thin films of chalcogenides such as CIGS and CZTS.

Table 2 shows the parameters used in the spray-pyrolysis deposition of absorber layers of copper-chalcogenide thin film from aqueous precursor solutions and the parameters obtained in the solar cells. As can be seen, most of the chalcogenide films are deposited using pneumatic atomizer, this reflects the simplicity and cost–benefit of this type of system. Other points to note are the prevalence of the use of N_2 as a carrier gas, and the flow rate that varies between 0.25 and 3 ml min^{-1}.

Mo-coated glass is among the most used substrates for depositing chalcogenides for PV applications. However, the preheating temperatures of the substrates required in the spray-pyrolysis process are high enough ($T \geq 200°C$) for the formation of Mo oxides. The oxide layer can lead to loss of charge carrier collection efficiency and worsen the adhesion of the film that will be deposited later. The solution found by Ho and colleagues [31] was to deposit the CIS layer with a gradual increase in substrate temperature from 120°C to 300–350°C. The XRD patterns shown in **Figure 3a** indicate that in the initial phase of deposition (red line) with temperatures between 120 and 150°C, in the intermediate phase with temperatures

	Deposition parameters						Solar cell parameters				
Absorber	Atomizer	Gas carrier	ultrasonic frequency (Hz)	Rate flow (mL min^{-1})	Substrate temperature (°C)	Atomization gas pressure (MPa)	V_{OC} (V)	J_{SC} ($mA\ cm^{-2}$)	FF (%)	PCE (%)	Ref.
CIGSSe/CdS	PSP	N_2		—	300–350	0.4	0.62	24.29	69.84	10.55	[35]
CZTS/CdS	PSP	Ar		3	220	0.2	0.41	10.07	33.0	1.4	[41]
$CZTS/In_2S_3$	PSP	N_2		1.2	450	0.1	0.43	28.27	47.07	5.54	[42]
CIGSSe/CdS	USP				430		0.64	27.43	61.93	10.89	[43]
ACZTSSe/CdS	PSP				380		0.67	18.5	57.6	7.14	[44]
CIGSSe/CdS	PSP	N_2		3	280	0.4	0.37	27.3	50.6	5.14	[45]
CCTS/CdS	USP		125	1	280		0.21	22.0	—	1.14	[46]
MCZTS/CdS	PSP	N_2			450		0.67	17.2	59.34	6.73	[47]
ACZTSSe/CdS	PSP	N_2			350		0.34	36.95	55.92	7.1	[48]
ACZTSSe/CdS	PSP	N_2		2	280	0.5	0.47	29.41	60.0	8.28	[24]

Table 2.
Atomizer parameters used in the spray-pyrolysis deposition of some copper-chalcogenides and the parameters obtained in solar cells.

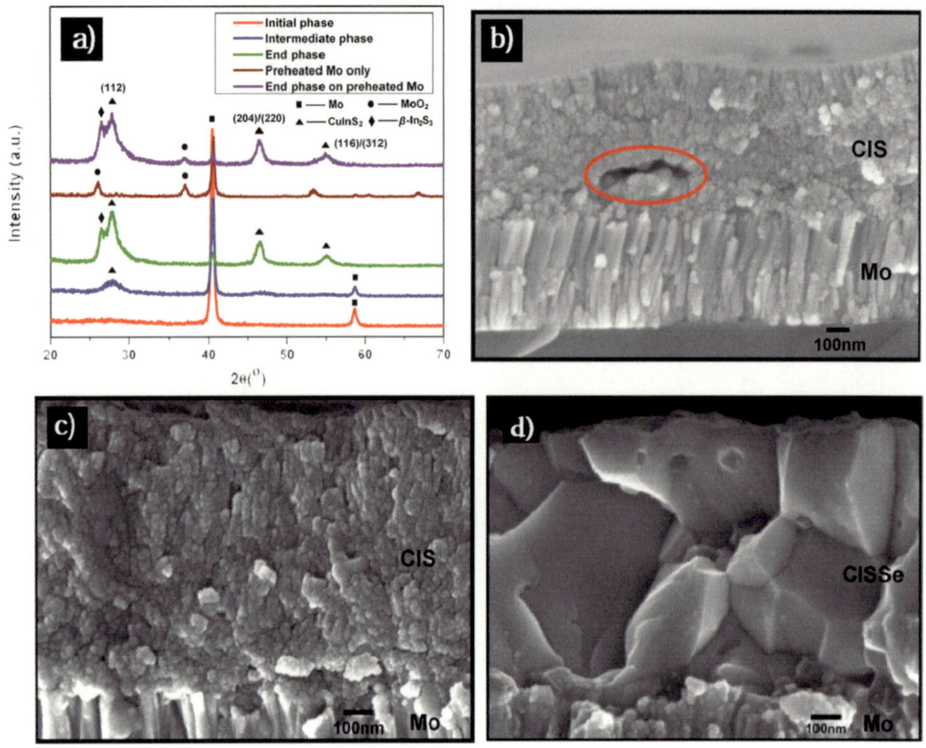

Figure 3.
(a) XRD patterns of the CIS films deposited by spray-pyrolysis at different substrate heat stage. (b) Cross-section SEM image of CIS film de posited on the preheated substrate. (c) Cross-section SEM image of CIS film deposited with gradual substrate heating. (d) CISSe film after heat treatment and selenization processes. Adapted with permission from [31]. Copyright (2014) American Chemical Society.

between 150 and 250°C (blue line) and the final phase with temperatures between 250 and 350° (green line) there were no characteristic peaks of MoO_2. On the other hand, in the films in which the Mo substrate was preheated between 300 and 350°C (violet line), peaks related to MoO_2 were found. **Figure 3b** and **c** show the scanning electron microscopy (SEM) images of films deposited with preheating and gradual heating, respectively. As can be seen, the CIS film deposited with preheating was less uniform and with pores at the CIS/Mo interface. The formation of Mo oxides is identified as the main cause of film malformation. Finally, most of the chalcogenides thin films deposited by spray-pyrolysis require a thermal treatment after their deposition, this is essential to increase the crystallinity, as shown in **Figure 3d**, and by means of sulfurization or selenization serves also for the stoichiometric improvement of S or Se deficient films.

4. Conclusions

Solution-processed chalcogenide thin film solar cells have already reached similar efficiencies to the ones prepared by using vacuum techniques. Spin-coating and spray-coating are inexpensive alternatives to fabricate highly efficient devices. The progress in the field since the hydrazine-based solution to deposit chalcogenide led to the development of environmentally friendly and low-cost molecular inks, nanocrystal inks and hybrid inks. Although solution-processed CIS and CZTS solar cells are equally to or more efficient than the vacuum-based devices, many improvements need to be done to put solution-based CIGS solar cells on top of the efficiency charts. Spin- and spray-coating are undoubtedly more advantageous for the process of solar cells fabrication, and more efforts are still needed to develop inks using environmentally friendly solvents, metals and chalcogen precursors. It is still needed to work on decreasing the residues in the films to eliminate any possible site of recombination. Optimization of these techniques will be essential to the scalability of the fabrication of the highly stable and highly efficient solar cells.

Acknowledgements

The authors acknowledge Fundação de Amparo à Pesquisa no Estado de São Paulo, FAPESP (grant #2017/15144-9, grant #2018/26005-2), FAPESP/CDMF (grant #2013/07296-2), FAPESP/SHELL (#2017/11986-5), FAPESP/GSK (#2014/50249-8); and FAPESP (grant #2019/26860-2).

Conflict of interest

The authors declare that there is no conflict of interest.

Author details

Marcos Antonio Santana Andrade Junior*, Hugo Leandro Sousa dos Santos, Mileny dos Santos Araujo, Arthur Corrado Salomão and Lucia Helena Mascaro
Department of Chemistry, Federal University of São Carlos, São Carlos, SP, Brazil

*Address all correspondence to: marcos_asaj@hotmail.com

References

[1] M. R. Golobostanfard and H. Abdizadeh, "All solution processable graded CIGS solar cells fabricated using electrophoretic deposition," *RSC Adv.*, vol. 6, no. 14, pp. 11903-11910, 2016, doi: 10.1039/c5ra26315h.

[2] M. A. Green, Y. Hishikawa, E. D. Dunlop, D. H. Levi, J. Hohl-Ebinger, and A. W. Y. Ho-Baillie, "Solar cell efficiency tables (version 52)," *Prog. Photovoltaics Res. Appl.*, vol. 26, no. 7, pp. 427-436, 2018, doi: 10.1002/pip.3040.

[3] T. M. Friedlmeier *et al.*, "High-efficiency Cu(In,Ga)Se 2 solar cells," *Thin Solid Films*, vol. 633, pp. 13-17, 2017, doi: 10.1016/j.tsf.2016.08.021.

[4] P. U. Londhe, A. B. Rohom, R. Fernandes, D. C. Kothari, and N. B. Chaure, "Development of Superstrate CuInGaSe2Thin Film Solar Cells with Low-Cost Electrochemical Route from Nonaqueous Bath," *ACS Sustain. Chem. Eng.*, vol. 6, no. 4, pp. 4987-4995, 2018, doi: 10.1021/acssuschemeng.7b04615.

[5] T. Todorov and D. B. Mitzi, "Direct liquid coating of chalcopyrite light-absorbing layers for photovoltaic devices," *Eur. J. Inorg. Chem.*, no. 1, pp. 17-28, 2010, doi: 10.1002/ejic.200900837.

[6] M. Graetzel, R. a J. Janssen, D. B. Mitzi, and E. H. Sargent, "Materials interface engineering for solution-processed photovoltaics.," *Nature*, vol. 488, no. 7411, pp. 304-12, Aug. 2012, doi: 10.1038/nature11476.

[7] D. B. Mitzi *et al.*, "A high-efficiency solution-deposited thin-film photovoltaic device," *Adv. Mater.*, vol. 20, no. 19, pp. 3657-3662, 2008, doi: 10.1002/adma.200800555.

[8] M. R. Golobostanfard, H. Abdizadeh, and A. Jannati, "Solution processable wurtzite CuInS2 inverted type solar cell," *Sol. Energy Mater. Sol. Cells*, vol. 164, no. January, pp. 1-6, 2017, doi: 10.1016/j.solmat.2017.02.004.

[9] J. A. Clark, A. Murray, J. M. Lee, T. S. Autrey, A. D. Collord, and H. W. Hillhouse, "Complexation Chemistry in N,N-Dimethylformamide-Based Molecular Inks for Chalcogenide Semiconductors and Photovoltaic Devices," *J. Am. Chem. Soc.*, vol. 141, no. 1, pp. 298-308, 2019, doi: 10.1021/jacs.8b09966.

[10] T. Schnabel, M. Seboui, and E. Ahlswede, "Evaluation of different metal salt solutions for the preparation of solar cells with wide-gap Cu2ZnGeSxSe4−x absorbers," *RSC Adv.*, vol. 7, no. 1, pp. 26-30, 2017, doi: 10.1039/c6ra23068g.

[11] A. Mishra, N. Bhatt, and A. K. Bajpai, *Nanostructured superhydrophobic coatings for solar panel applications*. Elsevier Inc., 2019.

[12] V. Slyke, "Organic Light-emitting Diodes," 1963, doi: 10.1016/B978-1-78548-158-1.50002-X.

[13] N. Guijarro *et al.*, "A Bottom-Up Approach toward All-Solution-Processed High-Efficiency Cu(In,Ga)S2 Photocathodes for Solar Water Splitting," *Adv. Energy Mater.*, vol. 6, no. 7, pp. 1-13, 2016, doi: 10.1002/aenm.201501949.

[14] N. Sahu, B. Parija, and S. Panigrahi, "Fundamental understanding and modeling of spin coating process: A review," *Indian J. Phys.*, vol. 83, no. 4, pp. 493-502, 2009, doi: 10.1007/s12648-009-0009-z.

[15] J. Ramanujam and U. P. Singh, "Copper indium gallium selenide based solar cells - A review," *Energy Environ. Sci.*, vol. 10, no. 6, pp. 1306-1319, 2017, doi: 10.1039/c7ee00826k.

[16] T. K. Todorov, O. Gunawan, T. Gokmen, and D. B. Mitzi, "Solution-processed Cu(In,Ga)(S,Se)2 absorber yielding a 15.2%," *Prog. PHOTOVOLTAICS Res. Appl.*, vol. 20, no. 6, pp. 1114-1129, 2012, doi: 10.1002/pip.

[17] C. J. Hibberd, E. Chassaing, W. Liu, D. B. Mitzi, D. Lincot, and A. N. Tiwari, "Non-vacuum methods for formation of Cu(In, Ga)(Se, S)2 thin film photovoltaic absorbers," *Prog. Photovoltaics Res. Appl.*, vol. 18, no. 6, pp. 434-452, 2010, doi: 10.1002/pip.914.

[18] D. B. Mitzi, O. Gunawan, T. K. Todorov, K. Wang, and S. Guha, "The path towards a high-performance solution-processed kesterite solar cell," *Sol. Energy Mater. Sol. Cells*, vol. 95, no. 6, pp. 1421-1436, 2011, doi: 10.1016/j.solmat.2010.11.028.

[19] S. Bag, O. Gunawan, T. Gokmen, Y. Zhu, and D. B. Mitzi, "Hydrazine-processed Ge-substituted CZTSe solar cells," *Chem. Mater.*, vol. 24, no. 23, pp. 4588-4593, 2012, doi: 10.1021/cm302881g.

[20] H. Azimi, Y. Hou, and C. J. Brabec, "Towards low-cost, environmentally friendly printed chalcopyrite and kesterite solar cells," *Energy Environ. Sci.*, vol. 7, no. 6, pp. 1829-1849, 2014, doi: 10.1039/C3EE43865A.

[21] W. Liu, D. B. Mitzi, M. Yuan, A. J. Kellock, S. Jay Chey, and O. Gunawan, "12% Efficiency CuIn(Se,S)2 photovoltaic device prepared using a hydrazine solution process," *Chem. Mater.*, vol. 22, no. 3, pp. 1010-1014, 2010, doi: 10.1021/cm901950q.

[22] A. R. Uhl, J. K. Katahara, and H. W. Hillhouse, "Molecular-ink route to 13.0% efficient low-bandgap CuIn(S,Se)2 and 14.7% efficient Cu(In,Ga)(S,Se)2 solar cells," *Energy Environ. Sci.*, vol. 9, no. 1, pp. 130-134, 2016, doi: 10.1039/c5ee02870a.

[23] M. H. Sayed, J. Schoneberg, J. Parisi, and L. Gütay, "Influence of silver incorporation on CZTSSe solar cells grown by spray pyrolysis," *Mater. Sci. Semicond. Process.*, vol. 76, no. November 2017, pp. 31-36, 2018, doi: 10.1016/j.mssp.2017.12.007.

[24] W. C. Huang, S. Y. Wei, C. H. Cai, W. H. Ho, and C. H. Lai, "The role of Ag in aqueous solution processed (Ag,Cu)2ZnSn(S,Se)4 kesterite solar cells: Antisite defect elimination and importance of Na passivation," *J. Mater. Chem. A*, vol. 6, no. 31, pp. 15170-15181, 2018, doi: 10.1039/c8ta02950d.

[25] M. G. Panthani *et al.*, "Nanocrystal ' Inks ' for Printable Photovoltaics," vol. 2, no. 7, pp. 16770-16777, 2008.

[26] H. W. Hillhouse and M. C. Beard, "Solar cells from colloidal nanocrystals: Fundamentals, materials, devices, and economics," *Curr. Opin. Colloid Interface Sci.*, vol. 14, no. 4, pp. 245-259, Aug. 2009, doi: 10.1016/j.cocis.2009.05.002.

[27] W. Wang, Y. W. Su, and C. H. Chang, "Inkjet printed chalcopyrite CuInxGa1-xSe2 thin film solar cells," *Sol. Energy Mater. Sol. Cells*, vol. 95, no. 9, pp. 2616-2620, 2011, doi: 10.1016/j.solmat.2011.05.011.

[28] W. Zhao, Y. Cui, and D. Pan, "Air-Stable, Low-Toxicity Precursors for CuIn(SeS)2 Solar Cells with 10.1% Efficiency," *Energy Technol.*, vol. 1, no. 2, pp. 131-134, 2013, doi: 10.1002/ente.201200044.

[29] J. A. Clark, A. R. Uhl, T. R. Martin, and H. W. Hillhouse, "Evolution of Morphology and Composition during Annealing and Selenization in Solution-Processed Cu2ZnSn(S,Se)4," *Chem. Mater.*, vol. 29, no. 21, pp. 9328-9339, 2017, doi: 10.1021/acs.chemmater.7b03313.

[30] C. Falcony, M. A. Aguilar-Frutis, and M. García-Hipólito, "Spray pyrolysis

technique; High-K dielectric films and luminescent materials: A review," *Micromachines*, vol. 9, no. 8, pp. 1-33, 2018, doi: 10.3390/mi9080414.

[31] J. C. W. Ho, T. Zhang, K. K. Lee, S. K. Batabyal, A. I. Y. Tok, and L. H. Wong, "Spray pyrolysis of CuIn(S,Se)2 solar cells with 5.9% efficiency: A method to prevent Mo oxidation in ambient atmosphere," *ACS Appl. Mater. Interfaces*, vol. 6, no. 9, pp. 6638-6643, 2014, doi: 10.1021/am500317m.

[32] R. A. Uhl *et al.*, "Non-vacuum deposition of Cu(In,Ga)Se2 absorber layers from binder free, alcohol solutions," *Prog. Photovolt Res. Appl.*, vol. 20, pp. 526-533, 2012, doi: 10.1002/pip.1246.

[33] M. Valdés, G. Santoro, and M. Vázquez, "Spray deposition of Cu2ZnSnS4 thin films," *J. Alloys Compd.*, vol. 585, pp. 776-782, 2014, doi: 10.1016/j.jallcom.2013.10.009.

[34] Y. Cai *et al.*, "Nanoparticle-induced grain growth of carbon-free solution-processed CuIn(S,Se)2 solar cell with 6% efficiency," *ACS Appl. Mater. Interfaces*, vol. 5, no. 5, pp. 1533-1537, 2013, doi: 10.1021/am303057z.

[35] M. A. Hossain *et al.*, "Synthesis of Cu(In,Ga)(S,Se)2 thin films using an aqueous spray-pyrolysis approach, and their solar cell efficiency of 10.5%," *J. Mater. Chem. A*, vol. 3, no. 8, pp. 4147-4154, 2015, doi: 10.1039/c4ta05783j.

[36] K. G. Deepa and N. Jampana, "Development of an automated ultrasonic spray pyrolysis system and the growth of Cu2ZnSnS4 thin films," *J. Anal. Appl. Pyrolysis*, vol. 117, pp. 141-146, 2016, doi: 10.1016/j.jaap.2015.12.004.

[37] S. Rahaman, M. A. Sunil, M. K. Singha, and K. Ghosh, "Temperature dependent growth of Cu2SnS3 thin films using ultrasonic spray pyrolysis for solar cell absorber layer and photocatalytic application," *Mater. Res. Express*, vol. 6, no. 10, 2019, doi: 10.1088/2053-1591/ab3928.

[38] D. Perednis and L. J. Gauckler, "Thin film deposition using spray pyrolysis," *J. Electroceramics*, vol. 14, no. 2, pp. 103-111, 2005, doi: 10.1007/s10832-005-0870-x.

[39] L. Filipovic *et al.*, "Methods of simulating thin film deposition using spray pyrolysis techniques," *Microelectron. Eng.*, vol. 117, pp. 57-66, 2014, doi: 10.1016/j.mee.2013.12.025.

[40] G. Blandenet, M. Court, and Y. Lagarde, "Thin layers deposited by the pyrosol process," *Thin Solid Films*, vol. 77, no. 1-3, pp. 81-90, 1981, doi: 10.1016/0040-6090(81)90362-X.

[41] M. E. Rodriguez *et al.*, "Pneumatically sprayed Cu2ZnSnS4 films under Ar and Ar-H2 atmosphere," *J. Phys. D. Appl. Phys.*, vol. 47, no. 24, 2014, doi: 10.1088/0022-3727/47/24/245101.

[42] D. B. Khadka, S. Y. Kim, and J. H. Kim, "A Nonvacuum Approach for Fabrication of Cu2ZnSnSe4/In2S3 Thin Film Solar Cell and Optoelectronic Characterization," *J. Phys. Chem. C*, vol. 119, no. 22, pp. 12226-12235, 2015, doi: 10.1021/acs.jpcc.5b03193.

[43] S. Kim, M. S. Mina, J. Lee, and J. H. Kim, "Sulfur-Alloying Effects on Cu(In,Ga)(S,Se)2 Solar Cell Fabricated Using Aqueous Spray Pyrolysis," *ACS Appl. Mater. Interfaces*, vol. 11, no. 49, pp. 45702-45708, 2019, doi: 10.1021/acsami.9b16192.

[44] T. H. Nguyen *et al.*, "Structural and Solar Cell Properties of a Ag-Containing Cu2ZnSnS4 Thin Film Derived from Spray Pyrolysis," *ACS Appl. Mater. Interfaces*, vol. 10, no. 6, pp. 5455-5463, 2018, doi: 10.1021/acsami.7b14929.

[45] X. Zeng *et al.*, "Cu2ZnSn(S,Se)4 kesterite solar cell with 5.1% efficiency using spray pyrolysis of aqueous precursor solution followed by selenization," *Sol. Energy Mater. Sol. Cells*, vol. 124, pp. 55-60, 2014, doi: 10.1016/j.solmat.2014.01.029.

[46] A. Tombak, T. Kilicoglu, and Y. S. Ocak, "Solar cells fabricated by spray pyrolysis deposited Cu2CdSnS4 thin films," *Renew. Energy*, vol. 146, pp. 1465-1470, 2020, doi: 10.1016/j.renene.2019.07.057.

[47] S. Lie *et al.*, "Improving Carrier-Transport Properties of CZTS by Mg Incorporation with Spray Pyrolysis," *ACS Appl. Mater. Interfaces*, vol. 11, no. 29, pp. 25824-25832, 2019, doi: 10.1021/acsami.9b05244.

[48] M. H. Sayed, J. Schoneberg, J. Parisi, and L. Gütay, "Influence of silver incorporation on CZTSSe solar cells grown by spray pyrolysis," *Mater. Sci. Semicond. Process.*, vol. 76, no. September 2017, pp. 31-36, 2018, doi: 10.1016/j.mssp.2017.12.007.

Section 5

VO_2 Polymorphs

Chapter 9

Thin Film Stabilization of Different VO_2 Polymorphs

Manish Kumar, Chirag Saharan and Sunita Rani

Abstract

In recent years, VO_2 has emerged as a popular candidate among the scientific community across the globe owing to its unique technological and fundamental aspects. VO_2 can exist in several polymorphs (such as: A, B, C, D, M_1, M_2, M_3, P, R and T) which offer a broad spectrum of functionalities suitable for numerous potential applications likewise smart windows, switching devices, memory materials, battery materials and so on. Each phase of VO_2 has specific physical and chemical properties. The device realization based on specific functionality call for stabilization of good quality single phase VO_2 thin films of desired polymorphs. Hence, the control on the growth of different VO_2 polymorphs in thin film form is very crucial. Different polymorphs of VO_2 can be stabilized by selecting the growth route, growth parameters and type of substrate etc. In this chapter, we present an overview of stabilization of the different phases of VO_2 in the thin film form and the identification of these phases mainly by X-ray diffraction and Raman spectroscopy techniques.

Keywords: thin film, VO_2, thermochromic, X-ray diffraction, Raman

1. Introduction

Thin film materials with 'smart' properties have attracted increasing attention in past few decades, as we move towards the smarter world [1]. This is driven by the fact that these materials react to the variation in parameters such as temperature, pressure, electric or magnetic fields etc. [2–13]. Vanadium dioxide (VO_2) is a well-known 'smart material' which is popular since the Morin' work in 1959 [14]. Its monoclinic M1 phase exhibits a metal–insulator transition (MIT) near room temperature, accompanied by larges changes in the structural, electronic and optical properties [15]. These distinctive features makes it attractive in smart windows, switching devices, memory materials and so on [16–18]. Being a strongly correlated electron system, VO_2 is equally attractive to condensed-matter physicists [19–22].

VO_2 can exhibit various polymorphic structures (such as: A, B, C, D, M_1, M_2, M_3, P, R and T), each having quite different physical and chemical properties [23–31]. Among these polymorphs, many are neither stable in ambient conditions nor can be easily synthesized. This happens because vanadium oxides can adopt a wide range of V:O ratios, resulting in different structural motifs. Phase space diagram (**Figure 1**) for the vanadium oxides indicates that there are more than 15 other stable vanadium oxides phases (like VO, V_2O_3, V_3O_5 etc.) and only a narrow window in phase space exist in which the pure semiconducting phase of VO_2 can be grown [32]. This narrow window strongly limits the synthesis of VO_2 either in the form of bulk crystals, thin films, or micro- and nanostructures. Nonetheless, different stoichiometric

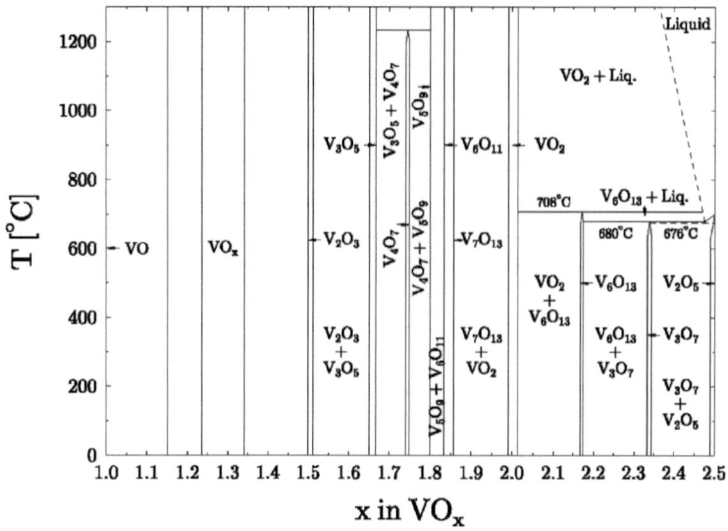

Figure 1.
Phase space diagram for the vanadium oxides. Note the narrow window within which stoichiometric VO_2 can be grown for x = 2.0 (reprinted from Ref. [32]).

Phase	Crystal structure (space group)	Lattice parameters				Comments and References
		a(Å)	b(Å)	c(Å)	β(°)	
VO_2 (A)	Tetragonal($P4_2/ncm$) (138)	8.43	8.43	7.68		[60]
VO_2 (B)	Monoclinic(C_2/m) (12)	12.03	3.69	6.42	106.6	[60]
VO_2 (C)	Tetragonal(I4/mnm) (139)	3.72	3.72	15.42		[24]
VO_2 (D)	Monoclinic(P2/c) (13)	4.59	5.68	4.91	89.3	[26]
VO_2 (P)	Orthorhombic(Pbnm) (62)	4.95	9.33	2.89		[28]
VO_2 (M_1)	Monoclinic($P2_1/c$) (14)	5.74	4.52	5.38	122.6	[61]
VO_2 (M_2)	Monoclinic(C2/m) (12)	9.08	5.76	4.53	91.3	[62]
VO_2 (M_3)	Monoclinic(P2/m) (10)	4.50	2.89	4.61	91.7	[62]
VO_2 (T)	Triclinic(P-1) (2)	9.06	5.77	4.52	91.4	[63]
VO_2 (R)	Tetragonal($P4_2/mnm$) (136)	4.55	4.55	2.86		[61]

Table 1.
The crystallography data for VO_2 polymorphs.

VO_2 polymorphs have been stabilized using techniques such as sputtering, pulsed laser deposition (PLD), sol–gel deposition, reactive evaporation and metal–organic chemical vapor deposition (MOCVD) etc. [15, 23, 25, 31, 33–38].

Koide and Takei appears to be the first to grow VO_2 thin films by chemical vapor deposition (CVD) technique in 1967 [39]. In their deposition method, fumes of vanadium oxychloride ($VOCl_3$) was carried by N_2 gas into the growth chamber and was hydrolyzed on the surface of rutile substrates to give epitaxial VO_2 films. In 1967, VO_2 thin films were also grown using reactive sputtering by Fuls et al. who made the films by reactive ion-beam sputtering of a vanadium target in an argon–oxygen atmosphere [40]. PLD emerged as a deposition technique for oxide superconductors in the late 1980s, and was first used to prepare VO_2 thin films by Borek et al. in 1993 [41]. Since then, consistent efforts have been made to grow thin films of various VO_2 polymorphs by using different deposition techniques/routes. Sputtering and PLD are the leading deposition techniques used to grow different VO_2 thin films polymorphs [42–46]. This is because of the ease with which one can play the deposition parameters in these techniques to stabilize thin films of various compounds [47–60].

In this chapter we will focus on the stabilization of thin film of six main VO_2 polymorphs: VO_2 (M_1), VO_2 (M_2), VO_2 (R), VO_2 (T), VO_2 (A) and VO_2 (B). But in passing it should be noted that VO_2 polymorphs likewise VO_2 (M_3), VO_2 (P), VO_2 (C) and VO_2 (D) have also been mostly studied in bulk and nanostructure form and reports are missing on thin film stabilization of these phases [24–29, 31]. Space group and lattice parameters of different VO_2 polymorphs known to us are tabulated in **Table 1**.

2. Thin film growth of different VO_2 polymorphs

2.1 VO_2 (M_1) and VO_2 (R) phase thin films

Monoclinic VO_2 (M_1) (a = 5.74 Å, b = 4.52 Å, c = 5.38 Å, β = 122.6°) with space group $P2_1/c$ is the most widely studied inorganic thermochromic material which is an insulator at room temperature. It shows a first-order MIT at 68°C with a concomitant structural transition into rutile tetragonal VO_2 (R) (a = b = 4.55 Å, c = 2.86 Å) having space group $P4_2/mnm$ [61]. Because of MIT and the associated huge changes in the structural, electronic and optical properties, VO_2 (M_1) and VO_2 (R) are attractive for applications in smart windows, switching devices, memory materials and so on [16, 17].

Figure 2 shows the structural arrangement of four different phases of VO_2 [64]. In the VO_2 (R) phase, the vanadium atoms are equally spaced along the rutile c axis (c_R), while in the VO_2 (M_1) phase, simultaneous dimerization and tilting in equivalent chains occur, leading to a zigzag pattern.

Highly oriented VO_2 (M_1) thin films on R-cut sapphire substrate were prepared by Borek et al. using PLD [41]. They ablated metallic vanadium target by a KrF pulsed excimer laser in an ultrahigh vacuum deposition chamber with Ar and O_2 (10:1) atmosphere of 100–200 mTorr, and a substrate temperature (T_s) ~ 500°C followed by 1 hour post deposition annealing in the same environment. Since then PLD was employed by number of groups to grow good quality VO_2 (M_1) thin films by varying the deposition parameters and post deposition treatment [44–46, 65]. Several other techniques such as sputtering, CVD, etc. were also employed to grow polycrystalline and epitaxial VO_2 (M_1) thin films on various substrates of different orientation [34, 42, 43, 66–69]. To date, most VO_2 (M_1) films have been grown on substrates such as sapphire (c-type, m-type, r-type and a-type), TiO_2, perovskite oxides, Si and Quartz. **Figure 3(a)** shows the grazing incidence X-ray diffraction (GIXRD) data of polycrystalline VO_2 (M_1) thin film by Kumar et al. which was

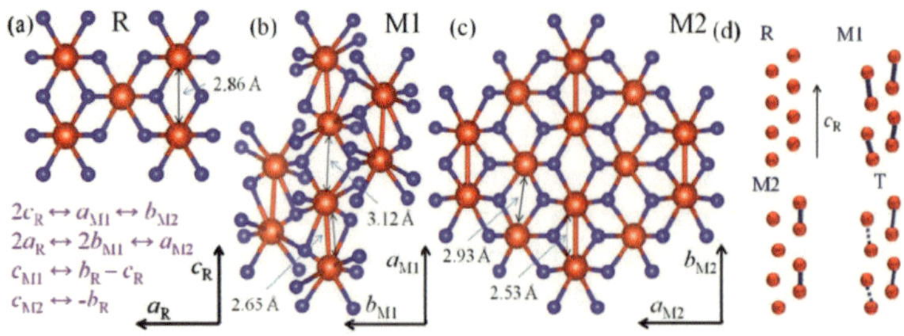

Figure 2.
The schematic structures for (a) rutile (R), (b) monoclinic M_1, and (c) M_2 phases of VO_2. Red and blue balls denote vanadium and oxygen atoms, respectively. (d) The arrangement of vanadium chains in the four phases without oxygen atoms (a-d reprinted from Ref. [64]).

grown on quartz substrate by sputtering VO_2 at room temperature and post deposition annealing at 500°C [69]. **Figure 3(b)–d** depict the X-ray diffraction (XRD) patterns of VO_2 (M1) thin film grown on TiO_2 and Al_2O_3 substrates of different orientation [46, 70].

VO_2 (R) is the high temperature phase of VO_2 (M_1). So, VO_2 (M_1) thin films generally transforms to VO_2 (R) phase when heated above the MIT temperature. Apart from this, thin films showing VO_2 (R) phase at room temperature can also be stabilized by strain, hydrogenation, oxygen vacancies and doping etc. [71–76]. Fan et al. reported the growth of ultrathin VO_2 (R) phase thin film on TiO_2 (002) substrate [71]. Y. Zhao et al. showed that hydrogenation can also lead to growth of VO_2 (R) phase thin film [72]. Very recently, Liang et al. described that increase in concentration of W dopant in $V_{1-x}W_xO_2$/Si thin films favors the growth of VO_2 (R) phase [73]. **Figure 4** shows the XRD patterns of VO_2 (R) phase thin films grown by different groups.

2.2 VO_2 (T) phase and VO_2 (M_2) phase thin films

VO_2 (T) phase and VO_2 (M_2) are known to be Mott-Hubbard type insulator which may find use in Mottronics and novel electronic transport applications [15, 18]. These phases are structurally different from VO_2 (M_1) and VO_2 (R) phase because of dissimilar types of vanadium chains and dimerization as shown in **Figure 2**. VO_2 (M_2) phase contains two distinct types of vanadium chains: one half of the vanadium atoms pair but do not tilt, while the other half are equidistant which tilts but do not pair. Triclinic phase i.e. VO_2 (T) phase can be thought of as an intermediate phase between VO_2 (M_1) and VO_2 (M_2) phases, having two types of inequivalent vanadium chains (or sublattices) in which the vanadium atoms are paired and tilted to different degrees. VO_2 (T) phase and VO_2 (M_2) are not as stable phase as VO_2 (M_1) and VO_2 (R). But, doping and/or strain can stabilize these phases [15, 35, 77]. Strelcov et al. presented a phase diagram which demonstrate the influence of chemical doping and uniaxial stress on the phase structure of VO_2 [35]. This phase diagram (**Figure 5(a)**) indicates that either of M_1, M_2, T, or R phase of VO_2 can exist depending on the type of dopant and/or stress. Majid et al. reported the Cr doping driven growth of VO_2 (T) phase thin films [15]. **Figure 5(b)** shows their XRD pattern of grown VO_2 (M_1) and VO_2 (T) phase thin films. Stress-induced VO_2 films with M_2 monoclinic phase stable at room temperature; were grown by Okimura et al. using inductively coupled plasma-assisted (ICP) reactive sputtering technique with various rf power fed to the coil for ICP (**Figure 5(c)**) at constant Ts of 450°C and at varying Ts, under constant rf power (**Figure 5(d)**) [77]. Apart from this work, there are not much reports on the growth of single phase VO_2 (M_2) thin films which are stable at room temperature. But, there are numerous reports on the evolution of intermediate M_2 phase in VO_2 thin films

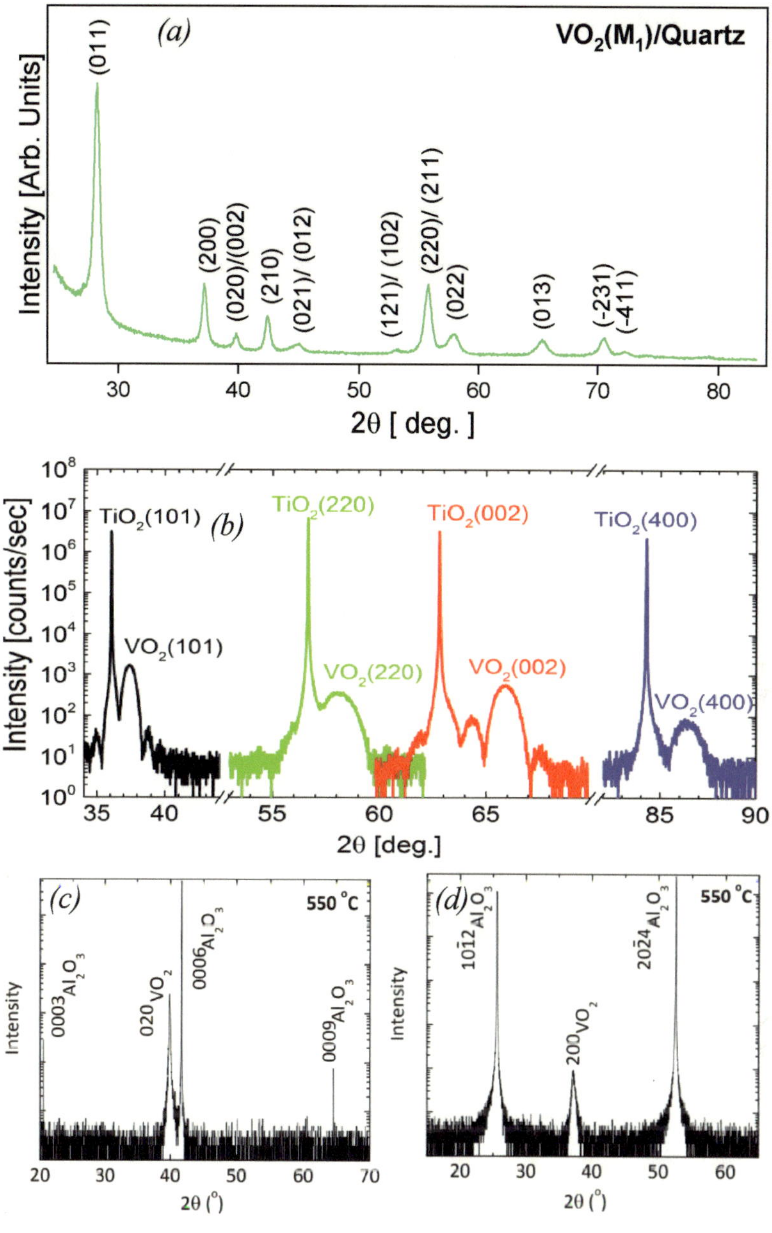

Figure 3.
(a) GIXRD data of VO_2 (M1) thin film prepared on quartz substrate [69]. XRD data of epitaxial VO_2 (M1) thin films grown on (b) TiO_2 substrates of different orientation (reprinted from Ref. [46]), (c) c-cut sapphire and (d) r-cut sapphire (c, d adopted from Ref. [70]).

during the monoclinic M_1 to rutile R transition [15, 69, 78–81]. This intermediate M2 phase in VO_2 thin film can be introduced by selecting the particular substrate temperature, doping, thickness etc. Kumar et al. witnessed the intermediate M2 phase temperature dependent XRD measurements across the MIT transition in polycrystalline VO_2 thin films grown on quartz substrate using sputtering technique followed by rapid thermal annealing at 530°C (**Figure 6(b)**) [69]. However, they have not observed the intermediate M_2 phase for films annealed at 500°C (**Figure 6(a)**). Majid et al. noticed the evolution of intermediate M_2 phase in temperature dependent Raman measurements of Cr doped VO_2 thin films during T ➔ R phase transition (**Figure 6(d)**) [15]. For undoped VO_2 thin films normal M_1➔R phase transition crossover was observed

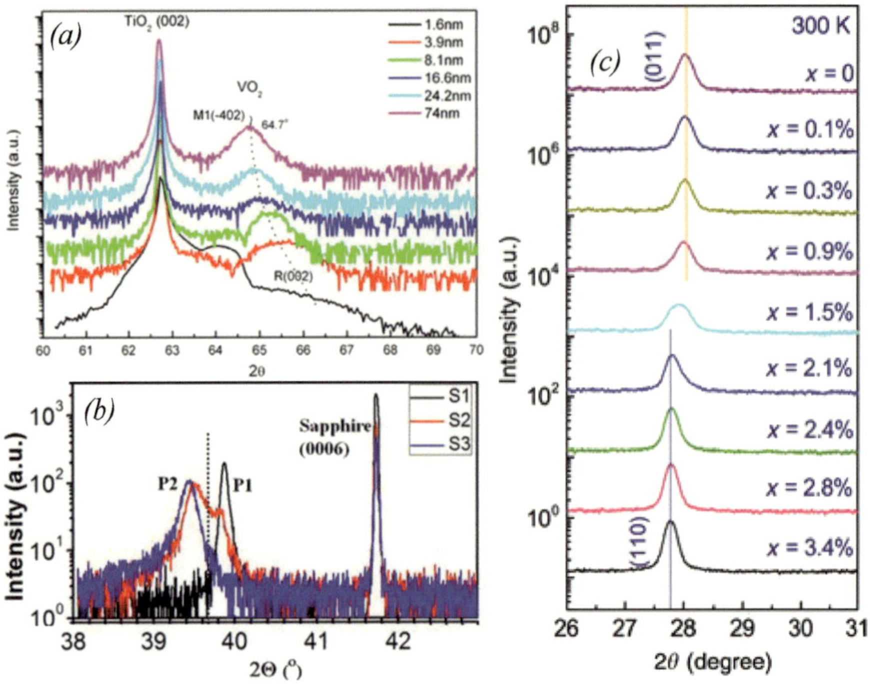

Figure 4.
(a) XRD profiles for thickness-dependence VO_2 films on TiO_2 substrate [Reprinted with permission from Fan et al [71]. Copyright (2014) American Chemical Society]. (b) XRD of pure (M_1 phase) and hydrogen-doping stabilized metallic (R phase) VO_2 thin films prepared on sapphire substrate (Reprinted from Ref. [72], with the permission of AIP Publishing). (c) Room temperature XRD of different $V_{1-x}W_xO_2$/Si thin films (adopted from Ref. [73]).

without signatures of intermediate M_2 phase °C (**Figure 6(c)**). Ji et al. stressed the role of microstructure on the M_1-M_2 phase transition in epitaxial VO_2 thin films of different thicknesses [78]. Their temperature dependent Raman measurement result on 90 nm and 150 nm thick VO_2 thin film sample are depicted in **Figure 6(e)** and **(f)** respectively. Azhan et al. also found intermediate M_2 phase in VO_2 thin films with large crystalline domains [79].

2.3 VO_2 (A) and VO_2 (B) phase thin films

The layered polymorphs VO_2 (A) and VO_2 (B) are important materials from science and technology perspective. VO_2 (B) has been long considered as a promising electrode material for Li ion batteries since the after report of Li et al. in 1994 [82]. It emerged as a promising cathode material owing to its layered structure and outstanding electrochemical performance [83, 84]. Also, it is important for the study of strong electronic correlations resulting from structure. On the other hand, VO_2 (A) phase is highly metastable and therefore the physical properties and the potential for technical applications have not been explored in detail. This phase is an intermediate phase between VO_2 (B) and VO_2 (R), and has a reversible phase transition at ~162°C [85, 86]. The crystal structure of VO_2 (A) and VO_2 (B) phase with possible epitaxial relation on $SrTiO_3$ substrate, are illustrated in **Figure 7(a)** and **(b)** respectively [23]. At room temperature, the metastable monoclinic VO_2 (B) adopts a structure derived from V_2O_5 and belongs to space group C2/m while VO_2 (A) adopts a tetragonal unit cell with a space group $P4_2/ncm$ [23]. Growth of single crystalline VO_2 (B) is very challenging due to the complex crystal structure. Similarly to VO_2 (B), the study of VO_2 (A) has so far been limited.

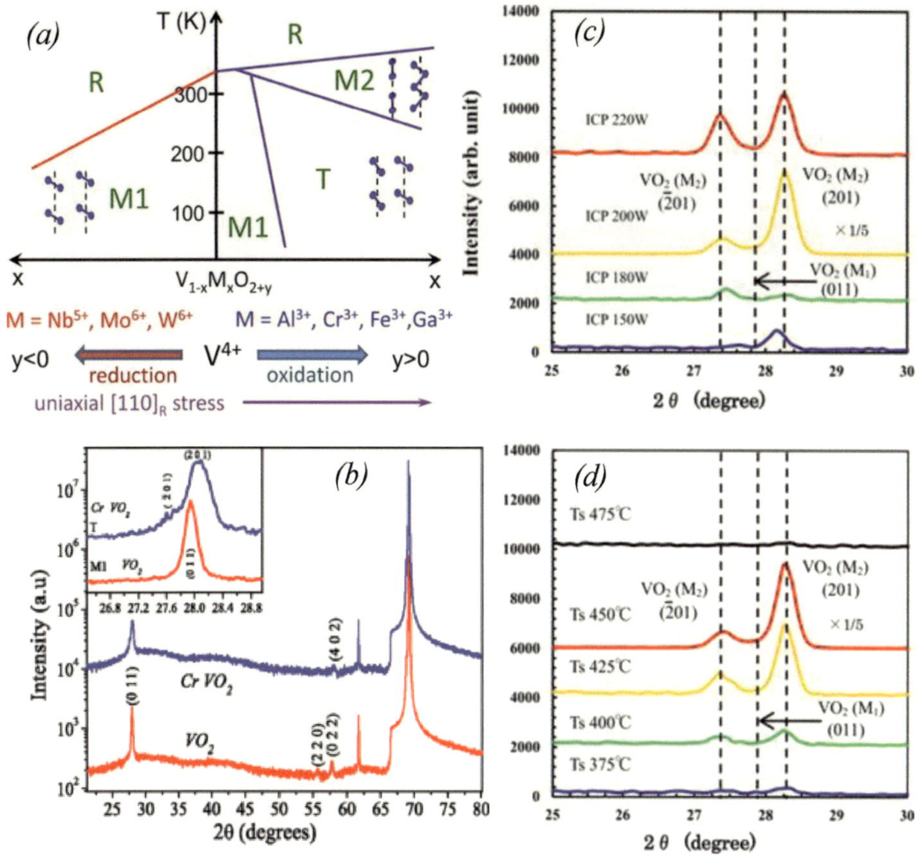

Figure 5.
(a) A temperature-composition phase diagram of VO_2, demonstrating the influence of chemical doping and uniaxial stress on the phase structure of VO_2 (reprinted with permission from Strelcov et al. [35]. Copyright 2012 American Chemical Society). (b), room-temperature XRD patterns of the pure (M1 phase) and Cr-doped (T phase) VO_2 thin films on the [001] Si substrate (adapted from Ref. [15]). (c and d) XRD patterns of the VO_2 films grown on quartz substrates with various RFpower fed to the coil for ICP, at constant Ts of 450°C and at varying Ts, under constant RF power (Reprinted from Ref. [77], with the permission of AIP Publishing).

Recently; several reports are focused on VO_2 (A) and VO_2 (B) phases in the form of bulk and nano-powders where annealing treatment causes them to revert to stable VO_2 (M_1) phase [25]. Chen et al. appears to be the first to report the growth of textured VO_2 (B) films with thickness only <25 nm on $SrTiO_3$ (001) substrate [87].

The good mathing of the a – b plane of VO_2 (B) to that of (001)-oriented perovskites enables the epitaxial growth of phase-pure VO_2 (B) thin films on perovskite substrates, such as $SrTiO_3$ and $LaAlO_3$. Srivastava et al. successfully stablized the single phase VO_2 (B) and VO_2 (A) thin films by tuning the laser retation rate and oxygen partical pressure during PLD while keeping the constant substrate tempearture (T_s = 500°C) [23]. The XRD pattern of their grown films and the phase digram of used deposition parameters are shown in **Figure 7(c)** and **(d)** respectively. Lee et al. argued that a proper choice of T_s is crtical among the deposition parameters for the growth of VO_2 (A) and VO_2 (B) phase thin film on perovskite substrates [60]. They found that the thin films of these phases can reproducibly grow at T_s lower than 430°C only (**Figure 8(a)** and **(b)**). Morover, VO_2 (A) phase can also appear as an intermediate phase (**Figure 8(c)**) when annealing is carried out for VO_2 (B)→ VO_2 (R) conversion [60]. Wong et al. successfully synthesize thin

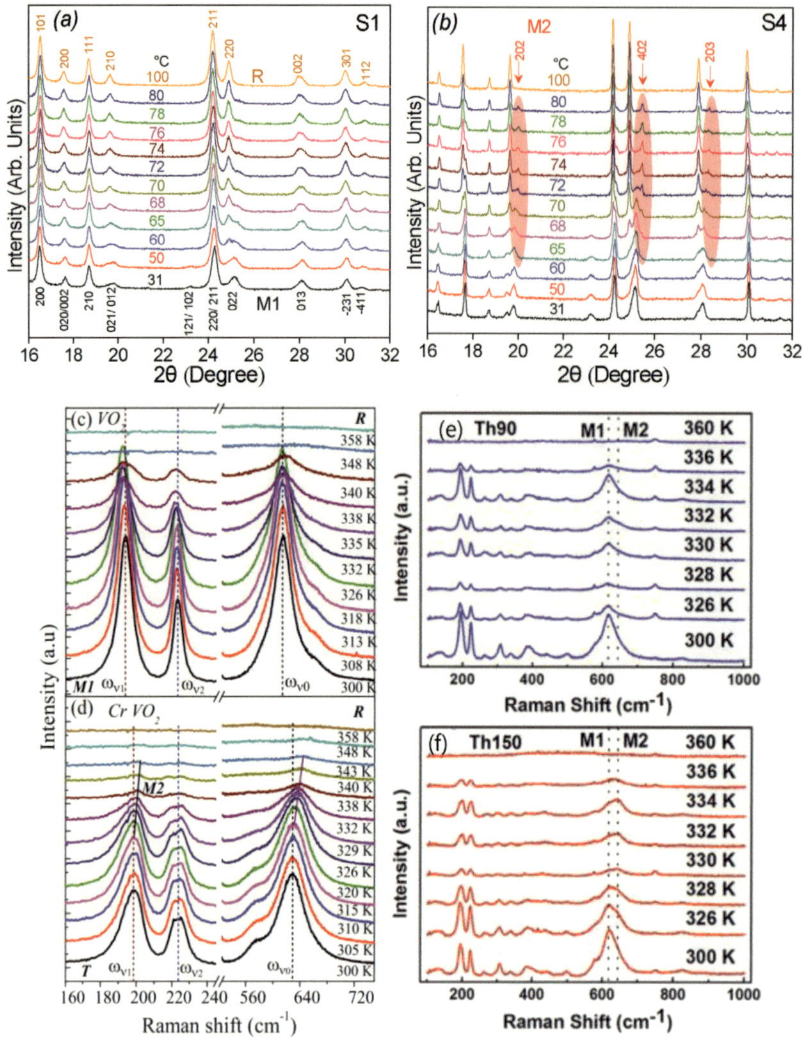

Figure 6.
Temperature dependence of XRD data (at X-ray wavelength (λ) = 0.0693 nm) during heating cycle for VO_2 thin film annealed at (a) 500°C and (b) 530°C (a,b adopted from Ref. [69]). Temperature-dependent Raman spectra of (c) pure and (d) Cr-doped VO_2 thin films collected in the cooling cycles (c, d adopted from Ref. [15]). Temperature dependent Raman spectra of (e) 90 nm and (f) 150 nm thick VO_2 thin film grown on Al_2O_3 substrate (e, f adopted from Ref. [78]).

films of the metastable VO_2 (B) polymorph on (001) $LaAlO_3$ at deposition temperature T_s = 325°C (**Figure 8(d)**) [70]. Very recently, Choi et al. grown epitaxial VO_2 (A) and VO_2 (B) thin films having tungsten doping were grown on (011)-oriented $SrTiO_3$ and 001)-oriented $LaAlO_3$ substrate respectively using PLD [88].

3. Conclusions

An overview of thin film stabilization of different VO_2 polymorphs i.e. VO_2 (M_1), VO_2 (M_2), VO_2 (R), VO_2 (T), VO_2 (A) and VO_2 (B) is presented in this chapter. It is understood that one can stabilize the thin film of a particular VO_2 polymorph by properly selecting the deposition technique, growth parameters, type of substrate and dopant etc.

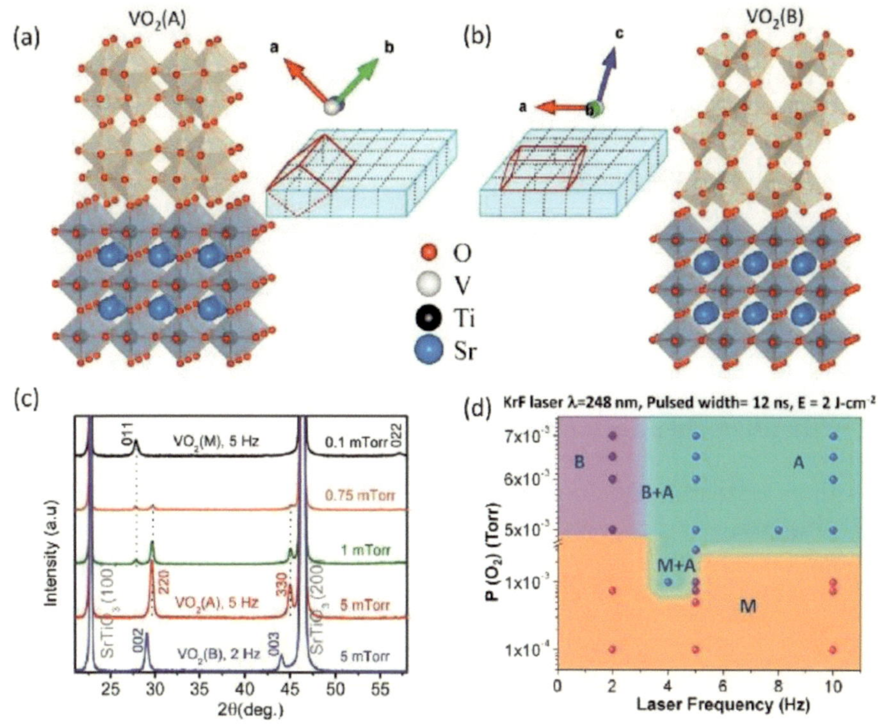

Figure 7.
The schematic crystal structure representation of (a) 220 orientated VO_2 (A), (b) 002 orientated VO_2 (B) grown on $SrTiO_3$ (100) substrate. (c) XRD patterns showing different phases for VO_2 thin films grown at various deposition parameters. (d) Phase diagram showing the role of laser frequency and oxygen pressure during pulsed laser deposition for different polymorphs of VO_2 thin films (a-d adopted from Ref. [23]).

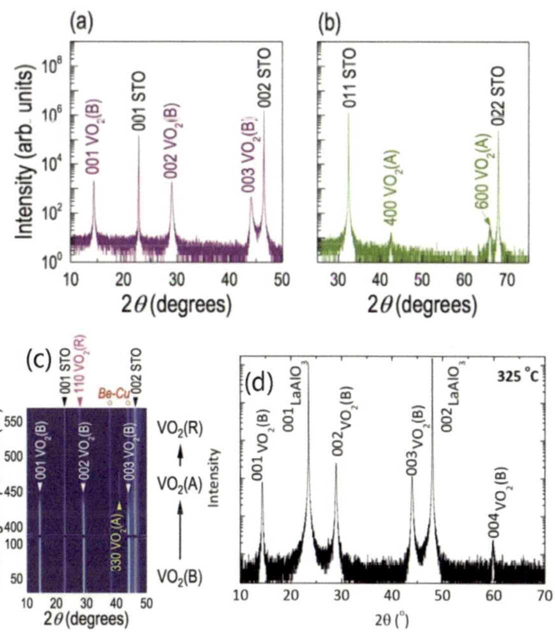

Figure 8.
XRD patterns of (a) VO_2 (B) and (b) VO_2 (A) thin film on $SrTiO_3$ (001) and (011) substrates respectively. (c) XRD during annealing of VO_2 (B)/STO sample (a-c adopted from Ref. [60]). (d) XRD scan of VO_2 (B) film grown on $LaAlO_3$ (001) substrate (adopted from Ref. [70]).

Acknowledgements

This work was supported by National Research Foundation of Korea (NRF) grant (Grant No. NRF-2015R1A5A1009962 and NRF-2019K1A3A7A09033398) funded by the Korean government. Authors also acknowledge the support from Pohang Accelerator Lab in Korea.

Conflict of interest

The authors declare no conflict of interest.

Notes/thanks/other declarations

Authors are thankful to the publisher for waive off the article processing charges of the chapter.

Author details

Manish Kumar[1*], Chirag Saharan[2] and Sunita Rani[1]

1 Pohang Accelerator Laboratory, Pohang University of Science and Technology, Pohang, 37673, South Korea

2 Department of Physics, Deenbandhu Chhotu Ram University of Science and Technology, Murthal, Sonepat (Haryana) - 131039, India

*Address all correspondence to: manish@postech.ac.kr

References

[1] Shahinpoor M, editor. Fundamentals of Smart Materials. 1st ed. Cambridge; Royal Society of Chemistry; 2020.338 p. ISBN: 9781782626459

[2] Kumar M, Phase DM, Choudhary RJ, Lee HH. Structure and functionalities of manganite/cuprate thin film. Current Applied Physics. 2018;**18S**:33-36. DOI: 10.1016/j.cap.2017.11.009

[3] Kumar M, Choudhary RJ, Shukla DK, Phase DM. Metastable magnetic state and magnetotransport in disordered manganite thin film. Journal of Applied Physics. 2014;**115**:163904. DOI: 10.1063/1.4873300

[4] Dagotto E. Complexity in strongly correlated electronic systems. Science. 2005;**309**:257. DOI: 10.1126/science.1107559

[5] Kumar M, Choudhary RJ, Phase DM. Valence band structure of $YMnO_3$ and the spin orbit coupling. Applied Physics Letters. 2013;**102**:182902. DOI: 10.1063/1.4804618

[6] Kumar M, Choudhary RJ, Phase DM. Magnetic and electronic properties of $La_{0.7}Ca_{0.3}MnO_3$/h-$YMnO_3$ bilayer. Journal of Vacuum Science and Technology A. 2016;**34**:021506. DOI: 10.1116/1.4937356

[7] Panchal G, Choudhary RJ, Kumar M, Phase DM. Interfacial spin glass mediated spontaneous exchange bias effect in self-assembled $La_{0.7}Sr_{0.3}MnO_3$: NiO nanocomposite thin films. J. Alloy. Compd. 2019;**796**:196-202. DOI: 10.1016/j.jallcom.2019.05.033

[8] Kumar M, Phase DM, Choudhary RJ. Structural, ferroelectric and dielectric properties of multiferroic $YMnO_3$ synthesized via microwave assisted radiant hybrid sintering. Heliyon. 2019;**4**:e01691. DOI: 10.1016/j.heliyon.2019.e01691

[9] Kumar M, Phase DM, Choudhary RJ, Upadhayay SK, Reddy VR. Microwave assisted radiant hybrid sintering of $YMnO_3$ ceramic: Reduction of microcracking and leakage current. Ceramics International. 2018;**44**:8196. DOI: 10.1016/j.ceramint.2018.01.268

[10] Kumar M, Choudhary RJ, Phase DM. Metastable magnetic state and exchange bias training effect in Mn-rich $YMnO_3$ thin films. Journal of Physics D: Applied Physics. 2015;**48**:125003. DOI: 10.1088/0022-3727/48/12/125003

[11] Kumar M, Choudhary RJ, Shukla DK, Phase DM. Superspin glassy behaviour of $La_{0.7}Ca_{0.3}Mn_{0.85}Al_{0.15}O_3$ thin film. Journal of Applied Physics. 2014;**116**:033917. DOI: 10.1063/1.4890507

[12] Kumar M, Choudhary RJ, Phase DM. Structural and multiferroic properties of self-doped yttrium manganites $YMn_{1+X}O_3$. AIP Conf. Proc. 2015;**1661**:07005. DOI: 10.1063/1.4915383

[13] Devi V, Kumar M, Wadikar AD, Choudhary RJ, Phase DM, Joshi BC. Electronic and multiferroic properties of $Zn0_{.85}Mg_{0.15}O$ thin film. AIP Conf. Proc. 2015;**1665**:080065. DOI: 10.1063/1.4917969

[14] Morin FJ. Oxides which show a metal-to-insulator transition at the neel temperature. Physical Review Letters. 1959;**3**:34. DOI: 10.1103/PhysRevLett.3.34

[15] Majid SS, Shukla DK, Rahman F, Khan S, Gautam K, Ahad A, et al. Insulator-metal transitions in the T phase Cr doped and M1 phase undoped VO_2 thin films. Physical Review B. 2018;**98**:075152. DOI: 10.1103/PhysRevB.98.075152

[16] Liu K, Lee S, Yang S, Delaire O, Wu J. Recent progresses on physics and applications of vanadium dioxide. Materials Today. 2018;**21**:875. DOI: 10.1016/j.mattod.2018.03.029

[17] Yang Z, Ko C, Ramanathan S. Oxide electronics utilizing ultrafast metal-insulator transitions. Annual Review of Materials Research. 2011;**41**:337. DOI: 10.1146/annurev-matsci-062910-100347

[18] Zhou Y, Ramanathan S. Mott memory and neuromorphic devices. Proceedings of the IEEE. 2015;**103**:1289. DOI: 10.1109/jproc.2015.2431914

[19] Shao Z, Cao X, Luo H, Jin P. Recent progress in the phase-transition mechanism and modulation of vanadium dioxide materials. NPG Asia Materials. 2018;**10**:581. DOI: 10.1038/s41427-018-0061-2

[20] Haverkort MW, Hu Z, Tanaka A, Reichelt W, Streltsov SV, Korotin MA, et al. Orbital-assisted metal-insulator transition in VO_2. Physical Review Letters. 2015;**95**:196404. DOI: 10.1103/PhysRevLett.95.196404

[21] O'Callahan BT, Jones AC, Park JH, Cobden DH, Atkin JM, Raschke MB. Inhomogeneity of the ultrafast insulator-to-metal transition dynamics of VO2. Nature Communications. 2015;**6**:6849. DOI: 10.1038/ncomms7849

[22] Gray AX, Jeong J, Aetukuri NP, Granitzka Chen PZ, Kukreja R, Higley D, et al. Correlation-driven insulator-metal transition in near-ideal vanadium dioxide films. Physical Review Letters. 2016;**116**:1. DOI: 10.1103/PhysRevLett.116.116403

[23] Srivastava A, Rotella H, Saha S, Pal B, Kalon G, Mathew S, et al. Selective growth of single phase VO_2(A, B, and M) polymorph thin films. APL Materials. 2015;**3**:026101. DOI: 10.1063/1.4906880

[24] Hagrman D, Zubieta J, Warren CJ, Linda MM, Michael MJT, Robert CH. A new polymorph of VO_2 prepared by soft chemical methods. Journal of Solid State Chemistry. 1998;**138**:178. DOI: 10.1006/jssc.1997.7575

[25] Li M, Magdassi S, Gao Y, Long Y. Hydrothermal synthesis of VO_2 polymorphs: Advantages, challenges and prospects for the application of energy efficient smart windows. Small. 2017;**13**:1701147. DOI: 10.1002/smll.201701147

[26] Liu L, Cao F, Yao T, Xu Y, Zhou M, Qu B, et al. New-phase VO_2 micro/nanostructures: Investigation of phase transformation and magnetic property. New Journal of Chemistry. 2012;**36**:619. DOI: 10.1039/c1nj20798a

[27] Song ZD, Zhang LM, Xia F, Webster N, Song J, Liu B, et al. Controllable synthesis of VO_2(D) and their conversion to VO_2(M) nanostructures with thermochromic phase transition properties. Inorganic Chemistry Frontiers. 2016;**3**:1035. DOI: 10.1039/C6QI00102E

[28] Wu C, Hu Z, Wang W, Zhang M, Yang J, Xie Y. Synthetic paramontroseite VO_2 with good aqueous lithium–ion battery performance. Chemical Communications. 2008;(33):3891. DOI: 10.1039/B806009F

[29] Braham E, Andrews JL, Alivio TEG, Fleer NA, Banerjee S. Stabilization of a metastable tunnel-structured orthorhombic phase of VO_2 upon iridium doping. Phys. Status Solidi A-Appl. Mat. 2018;**215**:1700884. DOI: 10.1002/pssa.201700884

[30] Park JH, Coy JM, Kasirga TS, Huang C, Fei Z, Hunter S, et al. Measurement of a solid-state triple point at the metal–insulator transition in VO_2. Nature. 2013;**500**:431. DOI: 10.1038/nature12425

[31] Galy J, Miehe G. Ab initio structures of (M2) and (M3) VO2 high pressure phases. Solid State Sciences. 1999;**1**:433. DOI: 10.1016/S1293-2558(00)80096-5

[32] Katzke H, Toledano P, Depmeier W. Physical Review B. 2003;**68**:024109. DOI: 10.1103/PhysRevB.68.024109

[33] MacChesney JB, Potter JF, Guggenheim HJ. Preparation and properties of vanadium dioxide films. Journal of the Electrochemical Society. 1968;**115**:52. DOI: 10.1149/1.2411002

[34] Kumar M, Singh JP, Chae KW, Park J, Lee HH. Annealing effect on phase transition and thermochromic properties of VO_2 thin films. Superlattices and Microstructures. 2020;**137**:106335. DOI: 10.1016/j.spmi.2019.106335

[35] Strelcov E, Tselev A, Ivanov I, Budai JD, Zhang J, Tischler JZ, et al. Doping-based stabilization of the M2 phase in free-standing VO_2 nanostructures at room temperature. Nano Letters. 2012;**12**:6198. DOI: 10.1021/nl303065h

[36] Sahana MB, Dharmaprakash MS, Shivashankar SA. Microstructure and properties of VO_2 thin films deposited by MOCVD from vanadyl acetylacetonate. Journal of Materials Chemistry. 2002;**12**:333. DOI: 10.1039/b106563g

[37] Warwick MEA, Binions R. Chemical vapour deposition of thermochromic vanadium dioxide thin films for energy efficient glazing. Journal of Solid State Chemistry. 2014;**214**:53. DOI: 10.1016/j.jssc.2013.10.040

[38] Seyfouri MM, Binions R. Sol-gel approaches to thermochromic vanadium dioxide coating for smart glazing application. Solar Energy Materials & Solar Cells. 2017;**159**:52. DOI: 10.1016/j.solmat.2016.08.035

[39] Koide S, Takei H. Epitaxial growth of VO2 single crystals and their anisotropic properties in electrical resistivities. Journal of the Physical Society of Japan. 1967;**22**:946. DOI: 10.1143/JPSJ.22.946

[40] Fuls EN, Hensler DH, Ross AR. Reactively sputtered vanadium dioxide thin films. Applied Physics Letters. 1967;**10**:199. DOI: 10.1063/1.1754909

[41] Borek M, Qian F, Nagabushnam V, Singh RK. Pulsed-laser deposition of oriented VO_2 thin films on R-cut sapphire substrates. Applied Physics Letters. 1993;**63**:3288. DOI: 10.1063/1.110177

[42] Manish K, Rani S, Lee HH. Thermochromic VO_2 thin films: Growth and characterization. AIP Conf. Proc. 2019;**2142**:080007. DOI: 10.1063/1.5122435

[43] Kumar M, Rani S, Lee HH. Effect of Ti:ZnO layer on the phase transition and optical properties of VO_2 film. Journal of the Korean Physical Society. 2019;**75**:519-522. DOI: 10.3938/jkps.75.519

[44] Kim DH, Kwok HS. Pulsed laser deposition of VO_2 thin films. Applied Physics Letters. 1994;**65**:3188. DOI: 10.1063/1.112476

[45] Émond N, Hendaoui A, Ibrahim A, Al-Naib I, Ozaki T, Chaker M. Transmission of reactive pulsed laser deposited VO_2 films in the THz domain. Applied Surface Science. 2016;**379**:377. DOI: 10.1016/j.apsusc.2016.04.018

[46] Jeong J, Aetukuri NB, Passarello D, Conradson SD, Samant MG. Parkin. Giant reversible, facet-dependent, structural changes in a correlated-electron insulator induced by ionic liquid gating. SSP. PNAS. 2015;**112**:1013. DOI: 10.1073/pnas.1419051112

[47] Kumar M, Choudhary RJ, Phase DM. Growth of different phases of yttrium manganese oxide thin films by pulsed laser deposition. AIP Conf. Proc. 2012;**1447**:655. DOI: 10.1063/1.4710173

[48] Devi V, Joshi BC, Kumar M, Choudhary RJ. Structural and optical properties of Cd and Mg doped zinc oxide thin films deposited by pulsed laser deposition. Journal of Physics: Conference Series. 2014;**534**:012047. DOI: 10.1088/1742-6596/534/1/012047

[49] Devi V, Kumar M, Kumar R, Joshi BC. Effect of substrate temperature and oxygen partial pressure on structural and optical properties of Mg doped ZnO thin films. Ceramics International. 2015;**41**:6269. DOI: 10.1016/j.ceramint.2015.01.049

[50] Devi V, Kumar M, Shukla DK, Choudhary RJ, Phase DM, Kumar R, et al. Structural, optical and electronic structure studies of Al doped ZnO thin films. Superlattices and Microstructures. 2015;**83**:431. DOI: 10.1016/j.spmi.2015.03.047

[51] Devi V, Kumar M, Choudhary RJ, Phase DM, Kumar R, Joshi BC. Band offset studies in pulse laser deposited $Zn_{1-x}Cd_xO$/ZnO hetero-junction. Journal of Applied Physics. 2015;**117**:225305. DOI: 10.1063/1.4922425

[52] Devi V, Kumar M, Kumar R, Singh A, Joshi BC. Band offset measurements in $Zn_{1-x}Sb_xO$/ ZnO hetero-junctions. J. Phys. D-Appl. Phys. 2015;**48**:335103. DOI: 10.1088/0022-3727/48/33/335103

[53] Devi V, Pandey H, Tripathi D, Kumar M, Joshi BC. Optical and electrical properties of pristine and Al doped ZnO thin films. AIP Conf. Proc. 2019;**2136**:040010. DOI: 10.1063/1.5120924

[54] Devi V, Kumar M, Choudhary RJ, Joshi BC. Structural and optical properties of $Zn_{1-x}Cd_xO$ thin films. AIP Conf. Proc. 2015;**1661**:110006. DOI: 10.1063/1.4915451

[55] Bhardwaj R, Kaur B, Singh JP, Kumar M, Lee HH, Kumar P, et al. Role of low energy transition metal ions in Interface formation in ZnO thin films and their effect on magnetic properties for Spintronics applications. Applied Surface Science. 2019;**479**:1021. DOI: 10.1016/j.apsusc.2019.02.107

[56] Kumar M, Singh JP, Chae KH, Lee HH. Structural and electronic properties of ZnO and Ti/Mn:ZnO flexible thin films. Journal of the Korean Physical Society. 2020;**77**:452. DOI: 10.3938/jkps.77.452

[57] Singh JP, Kumar M, Lim WC, Lee HH, Lee YM, Lee S, et al. MgO thin film growth on Si(001) by radio-frequency sputtering method. Journal of Nanoscience and Nanotechnology. 2020;**20**:7555. DOI: 10.1166/ jnn.2020.18613

[58] Kumar M, Singh JP, Chae KH, Kim JH, Lee HH. Structure, optical and electronic structure studies of Ti:ZnO thin films. J. Alloy. Compd. 2018;**759**:8. DOI: 10.1016/j. jallcom.2018.04.338

[59] Singh JP, Ji MJ, Kumar M, Lee IJ, Chae KH. Unveiling the nature of adsorbed species onto the surface of MgO thin films during prolonged annealing. J. Alloy. Compd. 2018;**748**:355. DOI: 10.1016/j. jallcom.2018.02.344

[60] Lee S, Ivanov IN, Keum JK, Lee HN. Epitaxial stabilization and phase instability of VO_2 polymorphs. Scientific Reports. 2016;**6**:19621. DOI: 10.1038/ srep19621

[61] Choi S, Chang SJ, Oh J, Jang HJ, Lee S. Electrical and optical properties of VO2 polymorphic films grown Epitaxially on Y-stabilized ZrO_2. Adv.

Electron. Mater. 2018;**4**:1700620. DOI: 10.1002/aelm.201700620

[62] Chamberland BL. New defect vanadium dioxide phases. Journal of Solid State Chemistry. 1973;**7**:377. DOI: 10.1016/0022-4596(73)90166-7

[63] Ghedira M, Vincent H, Marezio M, Launay JC. Structural aspects of the metal-insulator transitions in V0.985Al0.015O2. Journal of Solid State Chemistry. 1977;**22**:423. DOI: 10.1016/0022-4596(77)90020-2

[64] Basu R, Srihari V, Sardar M, Srivastava SK, Bera S, Dhara S. Probing phase transition in VO_2 with the novel observation of low-frequency collective spin excitation. Scientific Reports. 2020;**10**:1977. DOI: 10.1038/s41598-020-58813-x

[65] Yang TH, Aggarwal R, Gupta A, Zhou H, Narayan RJ, Narayan J. Semiconductor-metal transition characteristics of VO_2 thin films grown on c- and r-sapphire substrates. Journal of Applied Physics. 2010;**107**:053514. DOI: 10.1063/1.3327241

[66] Wong FJ, Zhou Y, Ramanathan S. Epitaxial variants of VO_2 thin films on complex oxide single crystal substrates with 3m surface symmetry. Journal of Crystal Growth. 2013;**364**:74. DOI: 10.1016/j.jcrysgro.2012.11.054

[67] Zhang H, Zhang L, Mukherjee D, Zheng Y, Haislmaier R, Alem N, et al. Wafer-scale growth of VO2 thin films using a combinatorial approach. Nature Communications. 2015;**6**:8475. DOI: 10.1038/ncomms9475

[68] Shao Z, Wang L, Chang T, Xu F, Sun G, Jin P, et al. Controllable phase-transition temperature upon strain release in VO_2/MgF_2 epitaxial films. Journal of Applied Physics. 2020;**128**:045303. DOI: 10.1063/5.0011423

[69] Kumar M, Rani S, Singh JP, Chae KW, Kim Y, Park J, et al. Structural phase control and thermochromic modulation of VO2 thin films by post thermal annealing. Applied Surface Science. 2020;**529**:147093. DOI: 10.1016/j.apsusc.2020.147093

[70] Wong FJ, Ramanathan S. Synthesis of epitaxial rutile-type VO_2 and VO_2 (B) polymorph films. Proc. of SPIE. 2014;**8987**:89870W. DOI: 10.1117/12.2044055

[71] Fan LL, Chen S, Luo ZL, Liu QH, Wu YF, Song L, et al. Strain dynamics of ultrathin VO2 film grown on TiO2 (001) and the associated phase transition modulation. Nano Letters. 2014;**14**:4036. DOI: 10.1021/nl501480f

[72] Zhao Y, Karaoglan-Bebek G, Pan X, Holtz M, Bernussi AA, Fan Z. Hydrogen-doping stabilized metallic VO2 (R) thin films and their application to suppress Fabry-Perot resonances in the terahertz regime. Applied Physics Letters. 2014;**104**:241901. DOI: 10.1063/1.4884077

[73] Liang YG, Lee S, Yu HS, Zhang HR, Liang YJ, Zavalij PY, et al. Tuning the hysteresis of a metal-insulator transition via lattice compatibility. Nature Communications. 2020;**11**:3539. DOI: 10.1038/s41467-020-17351-w

[74] Yoon H, Choi M, Lim T, Kwon H, Ihm K, Kim J, et al. Reversible phase modulation and hydrogen storage in multivalent VO_2 epitaxial thin films. Nature Materials. 2016;**15**:1113. DOI: 10.1038/nmat4692

[75] Lee D, Kim H, Kim JW, Lee IJ, Kim Y, Yun H, et al. Hydrogen incorporation induced the octahedral symmetry variation in VO_2 films. Applied Surface Science. 2017;**396**:36. DOI: 10.1016/j.apsusc.2016.11.047

[76] Jeong J, Aetukuri N, Graf T, Schladt TD, Samant MG, Parkin SSP.

Suppression of metal-insulator transition in VO_2 by electric field-induced oxygen vacancy formation. Science. 2013;**339**:1402. DOI: 10.1126/science.1230512

[77] Okimura K, Watanabe T, Sakai J. Stress-induced VO_2 films with M_2 monoclinic phase stable at room temperature grown by inductively coupled plasma-assisted reactive sputtering. Journal of Applied Physics. 2012;**111**:073514. DOI: 10.1063/1.3700210

[78] Ji Y, Zhang Y, Gao M, Yuan Z, Xia Y, Jin C, et al. Role of microstructures on the M1-M2 phase transition in epitaxial VO_2 thin films. Scientific Reports. 2014;**4**:4854. DOI: 10.1038/srep04854

[79] Azhan NH, Su K, Okimura K, Zaghrioui M, Sakai J. Appearance of large crystalline domains in VO_2 films grown on sapphire (001) and their phase transition characteristics. Journal of Applied Physics. 2015;**117**:245314. DOI: 10.1063/1.4923223

[80] Sharma Y, Holt MV, Laanait N, Gao X, Ivanov IN, Collins L, et al. Competing phases in epitaxial vanadium dioxide at nanoscale. APL Materials. 2019;**7**:081127. DOI: 10.1063/1.5115784

[81] Pouget JP, Launois H, D'Haenens JP, Merenda P, Rice TM. Electron localization induced by uniaxial stress in pure VO_2. Physical Review Letters. 1975;**35**:873. DOI: 10.1103/PhysRevLett.35.873

[82] Li W, Dahn JR, Wainwright DS. Rechargeable lithium batteries with aqueous electrolytes. Science. 1994;**264**:1115. DOI: 10.1126/science.264.5162.1115

[83] Lee S, Sun XG, Lubimtsev AA, Gao X, Ganesh P, Ward TZ, et al. Persistent electrochemical performance in epitaxial VO_2 (B). Nano Letters. 2017;**17**:2229. DOI: 10.1021/acs.nanolett.6b04831

[84] Xia C, Lin Z, Zhou Y, Zhao C, Liang H, Rozier P, et al. Large intercalation Pseudocapacitance in 2D VO2(B): Breaking through the kinetic barrier. Adv. Mat. 2018;**30**:1803594. DOI: 10.1002/adma.201803594

[85] Oka Y, Sato S, Yao T, Yamamoto N. Crystal structures and transition mechanism of VO_2 (a). Journal of Solid State Chemistry. 1998;**141**:594. DOI: 10.1006/jssc.1998.8025

[86] Zhang S, Shang B, Yang J, Yan W, Wei S, Xie Y. From VO_2 (B) to VO_2 (a) nanobelts: First hydrothermal transformation, spectroscopic study and first principles calculation. Physical Chemistry Chemical Physics. 2011;**13**:15873. DOI: 10.1039/C1CP20838A

[87] Chen A, Bi Z, Zhang W, Jian J, Jia QX, Wang H. Textured metastable VO_2 (B) thin films on $SrTiO_3$ substrates with significantly enhanced conductivity. Applied Physics Letters. 2014;**104**:071909. DOI: 10.1063/1.4865898

[88] Choi S, Ahn G, Moon SJ, Lee S. Tunable resistivity of correlated VO_2(A) and VO_2(B) via tungsten doping. Scientific Reports. 2020;**10**:9721. DOI: 10.1038/s41598-020-66439-2

Section 6

Thin Film Deposition

Chapter 10

Thin Film Deposition: Solution Based Approach

Kashif Tufail Chaudhary

Abstract

The wet chemical processing opens the way to deposit thin film with the versatility and ease for a variety of materials. Liquid film deposition involves the application of a liquid precursor on a substrate which is then converted to the required coating material in a subsequent post-treatment step. Different non-vacuum solution based deposition techniques have been developed to grow thin films with high efficiency and functionality. Spin coating is one of an effective technique for thin film fabrication due to low cost, uniformity, less hazardous, and capability of easy scaling up. The typical process involves depositing a small amount of a fluid onto the center of a substrate and then spinning the substrate at high speed. Dip coating is another simple, cost effective route with feasibility to scale-up for commercial production. The dip coating process can be divided into three important technical stages, immersion, withdrawal and evaporation. The coating may be subjected to further heat treatment in order to burn out residual compounds and induce crystallization of the functional oxides. Spray coating is a promising technique to grow thin film in research and industry to prepare thin and thick films. It is an easy approach to fabricate thin film with uniform distribution at small scale from a few nanometers to micrometers in thickness. Inkjet printing is the emerging promising technique to develop large-scale, and flexible thin films. The inkjet printing process allow easy customization to grow variety of complex structures.

Keywords: thin film deposition, spin coating, dip coating, spray coating, inkjet printing

1. Introduction

This chapter includes the details of solution based routes to deposit thin films which includes spray coating, dip coating, spin coating and inkjet printing processing. The contribution of different experimental parameters such as solution viscosity, surface tension, droplet size, substrate material & temperature, nature of solution are discussed briefly.

2. Spin coating

Spin coating is a quick and common route to deposit thin films on substrates with primary advantage of ease to produce very uniform films. The solution of a specific material is spun at high speeds, the centripetal force and the surface tension of the liquid together create an even covering on the substrate. The excessive solvent is evaporated, and spin coating results in a thin film ranging from a few nanometers to a few microns in thickness. Spin coating technique is used to coat small substrates from a few mm square to a metre or more in diameter. The key advantage of spin coating technique is the simplicity and relative ease to set up the process, coupled with the thinness and uniformity.

Spin coating consists of three major stages, solution dispensing, rotation dominated thinning and solvent evaporation as shown in **Figure 1**. The rotation pulls the liquid (solution) coating into an even covering and then evaporates to leave the desired material on the substrate in an even distribution. The high spin speeds and the high airflow leads to fast drying, which in turn results in high consistency at both macroscopic and nano length scales. However, the fast drying times lead to lower performance for certain processes, which requires time to self-assemble and/or crystallize.

Spin coating also relatively low throughput process due to an inherently batch (single substrate) process compared to roll-to-roll processes. Despite these drawbacks, spin coating is usually the starting point and benchmark for most academic and industrial processes that require a thin and uniform coating. Spin coating process can be broadly divided into 4 main stages.

Deposition: The first step involves the solution dispensing onto the substrate surface. The centrifugal force pulls the solution across substrate surface, if the substrate is already spinning (dynamic spin deposition) or is set on spin after deposition (static spin deposition).

Spin up: Once the solution is translated to the substrate surface, the rotation disc is speed up to the required rotation speed, either immediately or by a gradual speed up steps. In this stage, the solution initially spins at a different rate than the substrate speed, but eventually the two rotation speeds match up once the drag force balances rotational accelerations leads to the formation of thin fluid layer.

Spin off: During spin off stage, the solution begins to turn into thin film dominated by viscous forces. As the solution is flung off the substrate, the thin film changes color due to interference effects and it indicates completion of dry process.

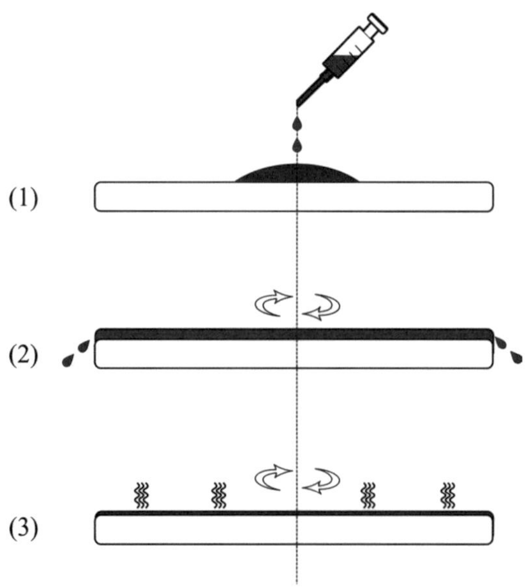

Figure 1.
Schematic of spin coating.

In certain cases, the edge effects are observed on the substrate due to formation of droplets at the edge and to be thrown off.

Evaporation: As fluid outflow ceased, the thinning process is dominated by evaporation of the solvent. The rate of solvent evaporation depends on the solvent vapor pressure, volatility, and ambient temperature.

All above processes are repeated several times to research desired film thickness. The solution casting and drying stages of are an integral and crucial part of the spin coating process, which contributes to the key process such as stacking/crystallization, phase separation and aggregation. Precise control of these processes is critical as the characteristics of deposited thin films not only depend upon morphology (thickness, uniformity) but also on deposition processing. In general, the spin speed speeds of >1000 rpm is recommended for industrial processing to ensure the high uniformity. However, spin speeds down to 200 rpm can be employed for laboratory scale deposition which might slow down the drying process but allow additional time for self-assembly.

In case of slow spinning, the solvent begins to evaporate as solvent is dispensed across the substrate and produces internal currents resulted in formation of highly ordered well assembled thin films by slowing down the evaporation rate. But, the coating across the surface is usually highly uneven with a typical "coffee staining" effect. However, it is possible to obtain a high level of nanoscale order by spinning at very lower speeds. Slow drop casting without rotation is a good way to deposit highly ordered thin films at the nanoscale but at the expense of uniformity.

In general, the thickness of a spin coated thin film is proportional to the inverse of the spin speed squared. However, the relation does not always apply and use to predict the film thickness without experimental data. Usually a test film is grown and measured. By using the data point(s) obtained from test film, spin thickness curve can be plotted with a reasonable accuracy. The spin speed can then be adjusted to obtain the right film thickness. The exact thickness of a deposited thin film is subjected to many factors as the material concentration and solvent evaporation rate which in turn depends upon the solvent viscosity, vapor pressure, temperature and local humidity etc. Therefore, the spin thickness curves for specific solution is commonly determined empirically [1].

The following parameters play very crucial role during spin coating process of thin film deposition.

Spin Speed: The spin speeds is one of the important parameter as it defines the range of thicknesses that can be achieved from a given solution. The speed affects the degree of centrifugal force applied to the liquid fluid as well as the velocity and characteristic turbulence of the air immediately above it. As the substrate starts spin, the dispensed solution flows radially, due to the action of centrifugal force, and the excess fluid spin off the substrate. The fluid continues thinning until the centrifugal force equals the viscosity. Although spin coating process is divided in three stages, however, they are not always very distinctly separate in time. In general, spin coating produces a uniform thin film relatively easily from about 1000 rpm upwards. A good quality thin film can be deposited at 500 or 600 rpm with precise control on spin speed. In general, the spin coating rpm might span a factor of ten (e.g. from 600 rpm to 6000 rpm) which in turn cause variation in film thickness. For instance, a solution which gives a film thickness of 10 nm at 6000 rpm may give a thickness of around 32 nm at 600 rpm. The dependence of thin film thickness upon the square root of spin speed has its own merits and demerits. The spin coating allows a precise control on the thickness of film but that can be achieved in a relatively narrow range for a given solution. The thickness of thin film for a given material/solvent combination also depends upon the maximum concentration of the material dissolved in the solvent. The thickness of >1 μm can be achieved for high solubility materials (100 mg/ml or higher). However, for less soluble materials (a few mg/ml), the maximum thickness might be limited to 20 nm or so and the thickness of a film is approximately linearly dependent upon the concentration of the material in the solution, however increase in the material concentrations affects the viscosity of the solution and thus a non-linear relationship develops. Films deposited at slow spin speed exhibit more edge-on orientation as compared to films at fast spin speed which exhibit more plane-on orientation, and the orientation varies smoothly between these limits [2].

Spin Coating Duration: In most standard spin coating processes, the substrate is kept spinning until the film is fully dried, which mainly depends upon the boiling point and vapor pressure of the solvent at ambient conditions. In general, 30 seconds are usually an adequate time for most of solvents. However, the drying may take considerably longer for higher boiling point solvents therefore solvents are most commonly used either as additives or with additional drying steps. The spin coating process involves a large number of variables that tend to cancel and average out during the spin process and allows sufficient time to manipulate deposition process. By combining the effects of fluid flow and solvent evaporation to leave a very thin layer of coating material on the substrate [3]. During evaporation process while spinning, the viscosity increases until equals the centrifugal force and spin time cannot move the solution anymore. At this point, the film thickness does not change significantly with increased spin time. This step can take from 10 seconds to 1 minute.

Solution Dispensing: Another important aspect is dispensing of the solution on substrate either in a static dispense deposition or a dynamic dispense deposition process. In a static dispense procedure, the solution is translated to the substrate while it is stationary. The entire substrate is covered with solution before rotation begins. The spinning is then started and raised up to required speed as fast as possible. The static dispense procedure usually covers the entire substrate (or at least all of the active part of the substrate) prior to initiate the spinning. The major challenge for a static dispense is the evaporation of solvent in the solution before the spin coating process begins. This issue is more pressing in case of high vapor pressure solvents. Therefore, the static dispense process gives lower substrate-to-substrate film uniformity as compared to dynamic dispense.

In a dynamic dispense, the substrate is first set on spinning and allowed to acquire desired speed and then the solution is dispensed at the center of the substrate. The centripetal force pulls the liquid away from the middle of substrate across the entire area before dries up. The dynamic dispense is more suitable due to precise controlling and better substrate-to-substrate variation. In this process, the solvent has less time to evaporate before the spinning start. Therefore, the ramp speed and dispense time is less critical as the substrate has been reached to the desired rpm. The dynamic dispense requires less solution in general although depends upon the wetting properties of the surface.

The major setback of the dynamic dispense is the incomplete substrate coverage in case of low spin speeds below 1000 rpm and viscous solutions due to insufficient centripetal force to pull the liquid across the surface. The lower rotation speed enhances the probability that the solution may be dispensed before the substrate has completed a full rotation. For the majority of spin coating above 1000 rpm, a dynamic dispense is recommended as standard process. A dynamic dispense process can be performed at speeds all the way down to around 500 rpm. However it becomes more difficult to get complete surface coverage.

The nature of solution: In spin coating process, the choice of the source compound and solvent must satisfy a number of essential conditions such as stable in air at room temperature, less or no toxic. In spin coating process different parameters of solution as viscosity, concentration, density, surface tension, evaporation rate of the solvent determines the spin speed and spin time to get desired thickness.

The nature of substrate: The nature of substrate material is also important parameter as it contributes towards the good coverage of substrate surface and quality of deposited thin film. Generally the substrate is chemical resistance to the coating solution and a heat resistance to withstand the subsequent curing or heat treatment. In certain cases, a surfactant is introduce into the formulation to facilitate complete substrate coverage during dispense or modifications are applied to the substrate to ensure good wetting. For instance, the substrate surface is changed from hydrophobic to hydrophilic depending on the nature of solution [1].

3. Dip coating

The dip coating is a facile, simple, the low cost and the high quality coating processing used for industrial as well as laboratory applications. The dip coating is commonly used for optical coatings such as in the production of automotive rear mirrors and large area antireflective coating for solar control glasses [4]. The dip coating process involves immersing of a substrate into the solution of coating materials and then withdraw the solution. The process can be defined as deposition of aqueous-based liquid phase onto the surface of substrate using a solution. Generally, the required material is dissolved in solutions and directly coated on the substrate surface, then the sedimentary (solvent) wet coating is evaporated to get dry film. The dip-coating process involves complex chemical and physical multivariable parameters. The film thickness and morphology depends on immersion time, withdrawal speed, dip-coating cycles, density and viscosity, surface tension, substrate surface and evaporation conditions of coating solutions. Photo-assisted dip-coating is used to control the evaporation process of coated solution and the irradiation effect facilitate the film disposition. To increase the uniformity and thickness of films, the multi-layered dip-coating is applied [5].

Dip coating consists of four basic steps as immersion, dwelling, withdrawal, and drying as shown in **Figure 2**. In first stage, the substrate is immersed into a solution to be deposited until it is completely covered with liquid. The substrate is then

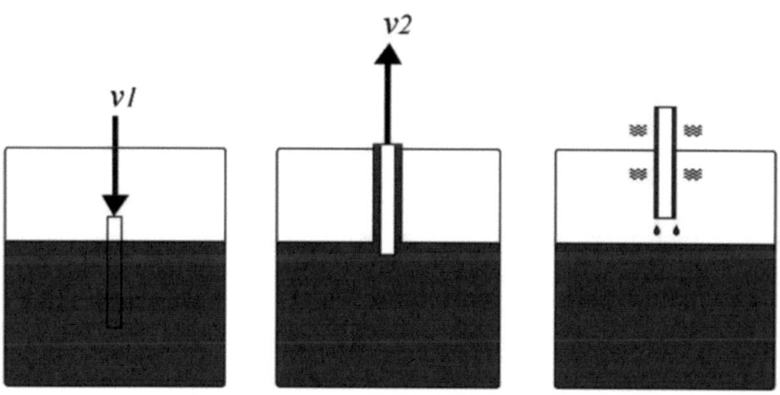

Figure 2.
Schematic of dip coating in lab.

withdrawn after a short interval. During the withdrawal process, a thin layer of the solution residues on the substrate surface. Once the substrate is fully withdrawn, the solution from the deposited film starts to evaporate and leaves behind a dry film. The deposited material undergoes a chemical or physical change.

The withdrawal and drying stages are critical stages to determine the properties of the deposited film. The withdrawal stage involves the interaction of different sets of forces. These forces are categorized into two categories, entraining forces and draining forces. Draining forces draw the liquid (solution) back to the bath from substrate surface. On the other hand, the entraining forces keeps the solution onto the substrate surface. The balance between these two sets of forces determines wet film thickness coated onto the substrate. During the withdrawal stage, the formation of the wet film can be broken into four regions as illustrated in **Figure 3**.

In the static meniscus region, the shape of the meniscus is determined by the balance of the hydrostatic and capillary pressures. Whereas the dynamic meniscus region occurs around the stagnation point where the entraining forces and draining forces are in equilibrium. In constant thickness zone, the wet film achieves a given thickness. The dynamic meniscus and the flow of solution in this region determine the wet film thickness. The transition between the dynamic and static meniscus happens within the boundary layer. Beyond the boundary layer, the draining forces are significantly higher as compared to the viscous forces and the balance between the capillary and hydrostatic pressure governs the meniscus shape.

Dip coating typically has three different stages for drying, drying front during coating, the falling rate period and the constant rate period. The simplest drying stages are the constant and the falling rate periods. The constant rate period occurs within the constant thickness zone which involves the evaporation of solvent at the surfaces of wet film across whole area. The only exception is the edges of the substrate, where the drying front occurs. Over the time after deposition, most of the solvents is evaporated from the wet film and left a gel-like film. This is when the

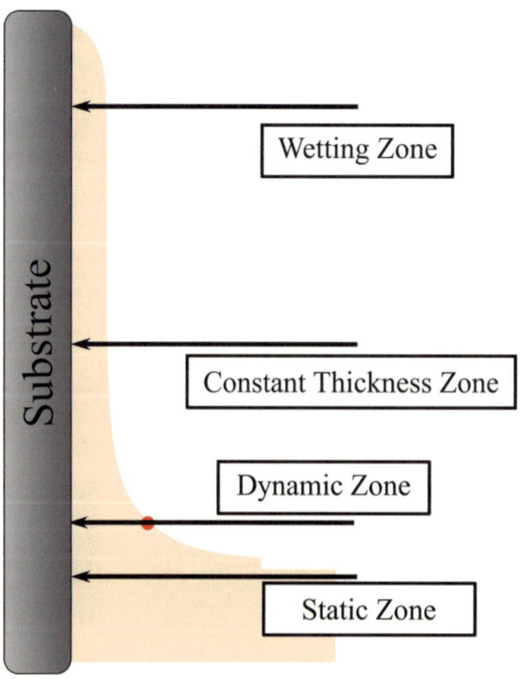

Figure 3.
Different regions during dip coating film formation.

falling rate period starts. In falling rate period, small amount of solvent left trapped within the gel material and the evaporation is determined by the diffusion of solvent towards the surface. The drying front appears at the interface between the wet film and substrate. Due to the large surface to volume ration, the evaporation occurs much faster at the surface which leads to the formation of a wet film with higher concentration. The more complex drying stage occurs at the drying front.

Several theoretical formulas were established to predict the thickness of deposited films, such as the Landau-Levich theory, via the following equation:

$$t_1 = 0.944 C_a^{1/6} \left(\frac{\eta U}{\rho g} \right)^{1/2} \tag{1}$$

where C_a is the capillary number and given by $C_a = \frac{\eta U}{\delta}$; η, δ, and ρ denote viscosity, surface tension, and the density of coating solutions, respectively; U is the withdrawal speed and g is the gravitational acceleration constant [5].

4. Spray coating

In recent years, spray coating has emerged as a viable approach for low-cost deposition of solution-processed thin films. Spray coating is a large-area, high-throughput, inexpensive, and industrially scalable process that can be used to create thin films of material which conform to the shape of the substrate. Spray coating involves ejecting fine liquid particles of smart materials by a jet stream of carrier gas onto the substrate as illustrated in **Figure 4**. Spray coating is a contact-free approach suitable for any substrate material and is particularly appropriate for low temperature processing [6]. The dynamics of spray droplet impingement on a substrate surface is a complex fluid mechanics problem subjected to different details, such as spreading, splashing, rebounding, coalescence and interaction with other droplets, drying phenomena, wetting/dewetting, and etc. Substrate properties such as roughness, permeability and surface energy also contribute significantly towards the droplet spreading and surface wetting. Spray coating on a permeable and rough surface hinders droplet spreading and increases the chance of splashing. Solution absorption by the substrate may also slow down droplet spreading. Therefore, droplet impact dynamics, such as droplet size and velocity, requires adjustment to enhance spreading and surface coverage on rough and permeable surface. The

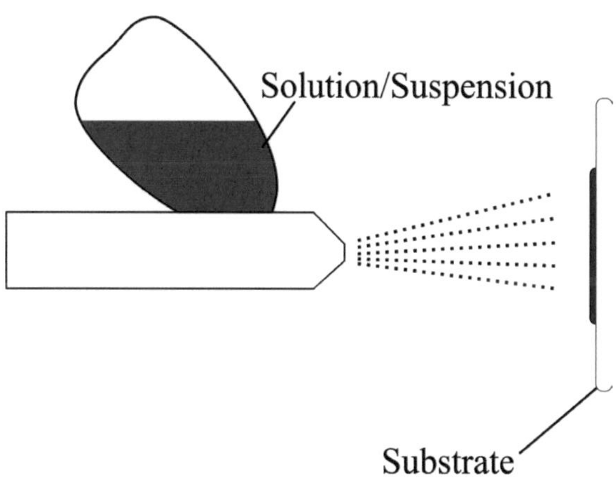

Figure 4.
Schematic of spray coating.

functionality of deposited thin film has a direct dependence on structure, morphology, roughness, and integrity of the stacked thin solid films [7].

The atomization process involves liquid breakup by the application of mechanical energy, which results in the production of a spray consist of micron-size drops. The solution characteristics such as liquid–vapor surface tension, viscosity, and density has direct influence on the atomization process. Atomization takes place when the dynamic pressure of an external force normally applied by a gas exceeds the internal pressure of the liquid droplet. These properties also affect morphological uniformity of deposited thin films. The increase in density hinder the movement of particulates and improve the morphological uniformity of a sprayed film [6].

Thin film formation by spray coating can be achieved by two different routes: one is by a drop-by-drop film formation approach, where the thin solid film is grown by impingement of a large number of individual droplets on the substrate surface that dry upon impact to cover the entire area. In second approach, the film is formed by immediate merging of impacted droplets and converted to a liquid film and to a solid film upon drying. The desired good quality solid film can be grown through drying of a liquid film if the spray process conditions such as the spray flow rate, and substrate temperature are optimized [7]. The influence of different experimental parameters on the thin film growth via spray coating is discussed as follow.

Effect of nozzle tip speed and solution precursor flow rate: The spray nozzle is not only deliver a mass of the solution to the substrate but also contributes towards the film characteristics as well by defining the number density, and size of the impinging droplets. The nozzle speed, nozzle height (distance from substrate), and flow rate (FR) (i.e., rate of liquid ejection from the spray nozzle) are closely related as they control the amount of liquid deposited per unit area [6, 7].

Effect of Number of Spray Cycles: Increase in the number of spray cycles considerably enhance the surface coverage and increases the film thickness. However, the influence of the number of spray cycles on film roughness is subjected to the substrate temperature. The application of the second spray cycle results in a smoother film, as the second layer generally covers up the cavities, pinholes, and defects of the first layer. This hypothesis is correct for low substrate temperatures. However at high substrate temperature, droplets of the first cycle are dried quickly and left behind a layer or film with irregular structure or grains. The forthcoming droplets from second spray cycle on rough layer at high substrate temperature dry quickly before they can spread and cover up the defective site and non-uniformities produced by the first layer [7].

Effect of Substrate Temperature: In general, the effect of substrate temperature on film characteristics depends on physical properties and evaporation rate of the precursor solution. Fabrication of a thin film at a high substrate temperature is determined by the balance between the heat transfer, droplet dynamics, and mass diffusion. The solvents are usually evaporated quickly, once the solution droplets impact on a hot substrate surface. Thus, there is a short time for droplet to spread and coalescence hinder surface wetting and complete coverage. At high substrate temperature, the precursor solution viscosity increases rapidly and further hinders droplet spreading. Despite the disadvantages of high substrate temperatures, the high temperature treatment can reduce the coffee ring effect and enhance high supersaturation with fast precipitation, which in turn can lead to the formation of well distributed and uniformly textured layers from each individual droplet. Therefore, a moderate temperature range can form a compromise between these two opposing effects (adequate and efficient coalescence and merging of impacted droplets on the substrate and timely evaporation and precipitation) may enhance the quality of the film [7].

Effect of Substrate Type: Regardless the substrate temperature, the coverage of films on regular surface is poor, due to limited spreading on rough substrates surface which leads to the higher film roughness, and the lower thickness is compared to the films made at the same condition but on the smooth surface. A large amount of the deposited precursor solution is consumed to fill the surface roughness and results in much less surface and higher non-uniformity and thickness. Implementation of a second spray cycle, however, can improve the coverage of coatings on rough substrate surface [7].

Effect of Substrate Vibration: To boost droplet coalescence and wetting process, the precursor solution is sprayed on substrate placed on an ultrasonic-assisted vibrating surface. A significant improvement in surface coverage, and films uniformity is observed when a vibrating substrate is employed. Substrate vibration improves coalescence and spreading of the impinging solution droplets, before being completely dried out. The impact of droplets onto a vibrating surface also results in further droplet breakup into several smaller droplets [7].

Distance between the nozzle and substrate: Some solvent boiling point is too low (63°C), the spray solution can dry before reaching the. It results in a thin film with relatively large density of pinholes and thickness variations, and the interface defect. Therefore, a proper nozzle-to-substrate distance is of great importance. The distance is determined by the solution properties, temperature of substrate and droplet size [8].

5. Inkjet printing

Inkjet printing is a relatively novel process compared to industrial printing and other thin film coating technologies. Inkjet printing involves delivering of a small volume of a fluid material, typically in the picoliter to nanoliter range, onto the substrate surface [9]. Inkjet printers comprises of three basic parts, the motion stage, control systems attached with the print heads, and vision system. The printer heads are connected directly to the cartridge filled with solutions. Inkjet printing systems typically require three mechanical degrees of freedom, one rotational (θ stage) and two translational (X and Y stages) to generate 2D patterns align with previously printed patterns and realize facilely stacked structures. The vision system is employed for substrate alignment and to observe ejected droplets in flight to monitor the droplets. Whereas, the control systems is used to optimize the stage and the printer head temperature, which can directly influence the substrate temperature and the dropping velocity, respectively [10].

Two different approaches are used to translate ink on the substrate surface: thermal and piezoelectric approach as depicted in **Figure 5**. The print cartridges consist of a series of tiny chambers and each chamber is attached with a heater over the nozzle. In case of thermal printing approach, a pulse of current is applied to the heater which leads to a rapid vaporization of the ink and build up pressure which push the droplet of ink out of nozzle [11]. The internal temperature creates a bubble in the cartridge and facilitates the single droplet formation of ink out of the nozzle via volume expansion. The negative pressure inside the cartridge after the droplet is ejected, draws new ink inside the reservoir [10]. In piezoelectric approach, the ink is ejected out from a nozzle through a sudden quasi-adiabatic reduction of the chamber volume via piezoelectric action. In the initial state, the ink in the printer head is in equilibrium state. Upon applying a voltage pulse signal to the piezoelectric element, the ink is ejected out due to the volume expansion inside the printer head. As the kinetic energy overcome the threshold, an in-flight droplet is generated and flights towards the target surface. In the sequentially applied opposite voltage, the ink in the printer head refills, and the whole process is repeated.

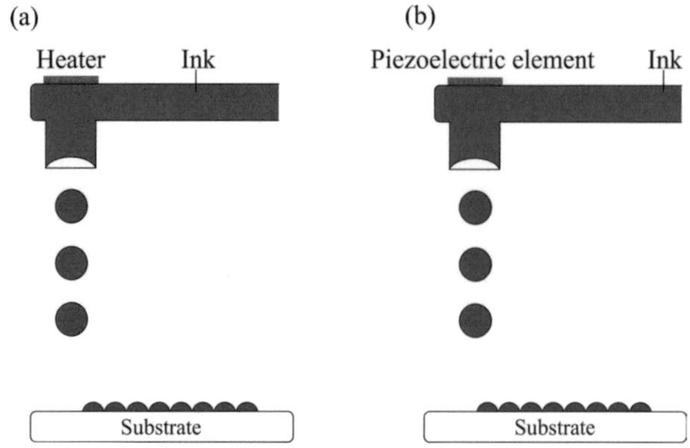

Figure 5.
Schematic of inkjet printing (a) thermal printing. (b) Piezoelectric printing.

Several thousand droplets are delivered to substrate in every second and pinter head also moves at the same time. Thermal printing is simple in design and low cost approach, however, it is confined to vaporizable inks to form bubble. The elevated operating temperature makes it inappropriate for polymer-based printing. Therefore, piezoelectric printing is widely used than thermal printing [12].

Ink viscosity and surface tension are crucial parameters during inkjet printing process. The low value of viscosity is required to allow the ink to fill the chamber as well as nozzle and the high surface tension enough to hold the ink in the nozzle without dripping [13]. To create droplets via thermal nozzles, the inks must be heat-compatible and sensitive to the volume contraction/expansion depending on the temperature. The piezoelectric nozzles contain a piezoelectric film along the wall of a reservoir. The deformation of the film generates the mechanical volume expansion in response to applied voltage pulses and as result the ink is ejected in response to the pressure generated by the piezoelectric element [10]. Typical inks used with a piezoelectric printing require a viscosity of 0.5–40 cp and a surface tension of 20–70 dyne/cm as the piezoelectric transducer only generates limited power, and printing high viscosity ink is difficult [14]. The piezoelectric nozzles have relatively better resolution and require lower temperature, which enables more precise operation to deposit thin film and do not suffer from ink degradation concerns and temperature-sensitive solvent choice. However thermal nozzles are typically less expensive and widely used in commercial printers [10].

Stable drop formation without satellite droplets after ejection from the nozzles is also important to obtain well-defined printed patterns on a substrate. The droplet velocity and volume are strongly depend on the pulse width and amplitude. The size of the droplets increases linearly with the size of the nozzle. However, the fine droplets generates high resolution and high surface morphology of final printed film. Usually nozzle size, droplets shape and fly direction are determined by the manufacture company and these parameters depend on the composition of the ink, especially on the solvent. Surface tension and viscosity are the primary physical properties that determine the shape and droplet-tail of in-flight droplets, and satellite droplet formation [10].

6. Summary

Spin coating, dip coating and spray coating are very common techniques and widely used to deposit thin films in research laboratories as well as industries. The spin coating is one of the important route for lab scale due high reproducibility and its suitability over a wide viscosity range. It is a quick and easy approach to grow uniform film at small scale from a few nanometers to a few micrometers in thickness. However, it is not suitable for scale up in industry due to its high material consumption and the restriction to large area. Dip coating is a commonly used method for mirror coating and dye processing as it can provide easy and fast deposition thin films over a large area. The key advantages is the large area processing and the thickness of film can be controlled by withdraw speed and viscosity of solution. Dip coating requires a high volume of coating solution and large tanks. Despite of long deposition life (several months) only about 20% of the solution can be used. The spray coating technique is able to access a broad spectrum of fluids, and offers the opportunity to tune the system to deposit any kinds of solution and obtain the desired film thickness. It is reproducible, and have great potential for large scale production. Inkjet printing is a powerful and cost-effective technique for deposition of liquid inks with high accuracy. The special characteristics offered by inkjet printing includes additive patterning, reducing materials consumption, non-contact deposition, low cost and capability of large area. Moreover, inkjet printing is capable of deposition a given material on a substrate that has pre-existing patterns, where contamination or damage of patterns would be induced with other deposition processes. However, to deposit materials on substrates, the solution or ink must be compatible with the print head and the viscosity should within a specific range.

Acknowledgements

The author would like to thank Ibnu Sina Institute for Scientific and Industrial Research (ISI-SIR), Universiti Teknologi Malaysia (UTM) for providing facilities. This research work has supported by Tier 1 Grant.

Author details

Kashif Tufail Chaudhary
Department of Physics, Universiti Teknologi Malaysia, Johor Bahru, Johor, Malaysia

*Address all correspondence to: kashif@utm.my

References

[1] Mouhamad, Y., Parvaneh Mokarian-Tabari, Nigel Clarke, R. A. L. Jones, and M. Geoghegan. "Dynamics of polymer film formation during spin coating." Journal of Applied Physics 116, no. 12 (2014): 123513.

[2] DeLongchamp, Dean M., Brandon M. Vogel, Youngsuk Jung, Marc C. Gurau, Curt A. Richter, Oleg A. Kirillov, Jan Obrzut et al. "Variations in semiconducting polymer microstructure and hole mobility with spin-coating speed." Chemistry of materials 17, no. 23 (2005): 5610-5612.

[3] Yimsiri, Piyachat, and Malcolm R. Mackley. "Spin and dip coating of light-emitting polymer solutions: Matching experiment with modelling." Chemical Engineering Science 61, no. 11 (2006): 3496-3505.

[4] Aegerter, Michel Andre, and Martin Mennig, eds. Sol-gel technologies for glass producers and users. Springer Science & Business Media, 2013.

[5] Tang, Xiaoning, and Xiong Yan. "Dip-coating for fibrous materials: mechanism, methods and applications." Journal of Sol-Gel Science and Technology 81, no. 2 (2017): 378-404.

[6] Carey, Tian, Chris Jones, Fred Le Moal, Davide Deganello, and Felice Torrisi. "Spray-Coating Thin Films on Three-Dimensional Surfaces for a Semitransparent Capacitive-Touch Device." ACS applied materials & interfaces 10, no. 23 (2018): 19948-19956.

[7] Zabihi, Fatemeh, and Morteza Eslamian. "Characteristics of thin films fabricated by spray coating on rough and permeable paper substrates." Journal of Coatings Technology and Research 12, no. 3 (2015): 489-503.

[8] Wang, Tao, Nicholas W. Scarratt, Hunan Yi, Alan DF Dunbar, Andrew J. Pearson, Darren C. Watters, Tom S. Glen et al. "Fabricating High Performance, Donor–Acceptor Copolymer Solar Cells by Spray-Coating in Air." Advanced Energy Materials 3, no. 4 (2013): 505-512.

[9] Cho, Ara, SeJin Ahn, Jae Ho Yun, Jihye Gwak, Seung Kyu Ahn, Keeshik Shin, Hyunjoon Song, and Kyung Hoon Yoon. "Non-vacuum processed CuInSe2 thin films fabricated with a hybrid ink." Solar energy materials and solar cells 109 (2013): 17-25.

[10] Chung, Seungjun, Kyungjune Cho, and Takhee Lee. "Recent progress in inkjet-printed thin-film transistors." Advanced science 6, no. 6 (2019): 1801445.

[11] Guo, Qijie, Grayson M. Ford, Rakesh Agrawal, and Hugh W. Hillhouse. "Ink formulation and low-temperature incorporation of sodium to yield 12% efficient Cu (In, Ga)(S, Se) 2 solar cells from sulfide nanocrystal inks." Progress in Photovoltaics: Research and Applications 21, no. 1 (2013): 64-71.

[12] Zhan, Zhaoyao, Jianing An, Yuefan Wei, and Hejun Du. "Inkjet-printed optoelectronics." Nanoscale 9, no. 3 (2017): 965-993.

[13] Calvert, Paul. "Inkjet printing for materials and devices." Chemistry of materials 13, no. 10 (2001): 3299-3305.

[14] Aegerter, Michel Andre, and Martin Mennig, eds. Sol-gel technologies for glass producers and users. Springer Science & Business Media, 2013.

Section 7

Sol-Gel Dip Coating Method

Chapter 11

TiO_2 Thin Films Produced by Ultrasonic-Agitation Assisted Sol-Gel for Propane Gas Sensor

Dwight Acosta-Najarro, Ismael A. Garduño-Wilches, Maria de la Luz Olvera and Arturo Maldonado

Abstract

TiO_2 films were deposited by the sol–gel dip coating method assisted by ultra sonic agitation, for their use as propane gas sensors. XRD anlaysis revealed the TiO_2 anatase phase for all samples. The film thickness was controlled with the number of the dip-coating immersion cycles, obtaining thicknesses from 10 to 130 nm and a growing rate of 18 nm per immersion. Surface morphology shows that films grow more compact and densely packed as the number of immersion cycles increase. Gas sensing studies indicate that work temperatures above 300°C are needed for sensing activation of the films. Gas sensors sensitivity as a function of propane concentration and film thickness revealed a two-regime behavior: for partial pressures of propane above 400 ppm the sensitivity is not dependent of the film thickness, whereas it is dependent on this parameter at lower partial pressures. Such behavior is attributed to a space charge region and a change in surface structure. For the low concentration regime, the maximum sensitivity was obtained for a film thickness of 110 nm or 5 inmersions.

Keywords: titanium dioxide, thin films, sol–gel, gas sensors

1. Introduction

Propane gas, found in natural gas and produced in the petroleum refining process among other alkanes, is widely used as combustible in the whole world. Propane is one of the main components of the liquefied petroleum gas, which is distributed to homes and buildings for water heating systems and cooking; it is also distributed by fuel stations to be used in transport vehicles, included public transport. Due to its low cost in the last years the use of natural gas as fuel has increased. Propane is used in power production plants as well as a precursor to chemicals such as isopropyl alcohol and silicon carbide [1, 2]. Because of propane economical importance, its increasing use as power source and its explosiveness in presence of oxygen, it is necessary to detect and control accurately its presence in public buildings, homes and industry. Such task implies the development of several sensing mechanisms, control valves and pipes. Because of this, several propane sensors have been developed and, among many others, those based in changes in electrical conductivity are some the most promising; however, there is still too much work to do when it comes to sense propane in more cheap, accurate and quick ways.

Metal oxides such as NiO, ZnO, SnO_2 and TiO_2 are widely used in several applications, in particular as gas sensors based in electrical conductivity modification; among them titanium dioxide is one of the favorites. Because of its large dielectric constant, its chemical stability, its durability, its biocompatibility and its band gap, titanium dioxide has been exploited to build capacitors for microelectronic systems [3–5], optical materials, including white paint and sunscreen, photocatalytic systems [6–8], and have given it, potential applications in medicine due to its antibacterial properties [9, 10]. TiO_2 and the metal oxides mentioned also changes its electrical resistance as a function of the chemical composition of its surrounding atmosphere, making them suitable for gas sensing [11, 12].

Titanium dioxide gas sensing properties have been widely studied and modified by doping it with metals and supported catalysts such as Pt [13–15]. Still, several efforts have to be done to produce more effective gas sensors: since they present disadvantages in response and recovering times, and in gas selectivity. Thus, studies on the response of these oxides must be realized in order to determine particular responses for each gas and improve, in this way their selectivity. Titanium dioxide response to CO, CH_4 and H_2 has been studied [16–19] and in some articles the response to alcohols such as ethanol, methanol and propanol has been reported [20–22]. In the literature, few works report about the response of the electrical conductivity of titanium dioxide, when it is exposed to propane gas [23].

Many synthesis routes have been used to obtain titanium oxide, either as a powder and as a thin film. Some are complex and expensive as is the case of sputtering and electron beam deposition [24, 25]. Some others are cheap and easy to work with, as is the case of spray pyrolysis, spin coating, dip coating and chemical bath deposition [26–29]. Also, new techniques or modified versions of the existing techniques are being explored, as is de case of the sol–gel ultrasonically assisted that has been used to obtain pure rutile phase powders and to obtain smaller grain sizes and larger surface areas when compared with the standard sol–gel technique [30, 31]. This last result is promising for gas sensing, since the sensor sensitivity depends directly on the surface area, it can be expected that more sensitive sensors can be made by synthesizing with ultrasonic assistance.

Therefore, in this work it TiO_2 films are obtained by the dip-coating method using the ultrasonic assistance during the crystal growth stage. The film thickness and work temperature are used as variables to explore the response to propane gas at different concentrations. The surface morphology is studied for different inmersion cycles and the crystalline structure is analyzed by x-ray diffraction.

2. Experimental

Titanium dioxide films were grown on glass substrates. Previous to deposition, substrates were cleaned in an ultrasonic bath with soap water, methanol and acetone, in an ultrasonic bath, and were dried under nitrogen gas flow. The start solution was prepared from two separated solutions (A and B). Solution A consisted of 1.04 ml of titanium isopropoxide dissolved in 10 ml of alcohol; while solution B consisted of 0.14 ml of deionized water dissolved in the same amount of alcohol. Once obtained, solution B was dropwise added to solution A while ultrasonically stirring, finally 0.4 ml of acetic acid was added to maintain a neutral ph. The solution was ultrasonically stirred for 10 min after both solutions were mixed.

A set of seven TiO_2 films with different thickness was produced by immersion of glass substrates in the starting solution. The films thicknesses was controlled with the number of immersion cycles, each cycle consisting of an immersion and extraction of the substrate in the solution at a speed of 0.122 cm/s. All samples were dried for 10 min in a furnace at 400°C after each immersion cycle. Once all cycles were carried out, samples were treated at 400°C for 3 hr. TiO_2 thin films with different thickness were obtained systematically after 1 to 7 immersion cycles.

Film thicknesses were determined using both a Bruker Dektak XT profilometer and a Filmetrics F20 analyzer. The F20 was operated from 380 to 1100 nm and an ultra clean glass substrate was used as a reference. The crystalline structure was determined by X-ray diffraction analysis using a Bruker D8 diffractometer with a CuK_α (1.54056 Å) wavelength, at an incidence angle of 1° and varying 2θ from 20° to 80°, and a recording time of 1.5 hrs to reduce noise. The X-ray diffraction patterns were interpreted in reference to the PDF card: 21-1272 from the ICCD data base. Samples surfaces were analyzed with a LV 5600 JEOL Scanning Probe Microscope operated in tapping mode.

Sensitivity tests were performed in a vacuum chamber equipped with a mechanical vacuum pump, a temperature-controlled sample holder, and a propane gas inlet; tungsten wires were used to establish a series connection of the sensor with a Keythley 2001 multimeter used to measure changes in sensors electrical resistance. Diagrams of the sensing chamber have been reported elsewhere [32]. Previous to start the sensing tests, two silver contacts were painted on films surface to allow a direct measurement of their electrical properties. To control samples temperature the sample holder is equipped with an electrical resistance as heater and a K-type thermocouple, both connected to a temperature controller. Sensors' temperature was set to take values of ambient, 100°C, 200°C and 300°C. At each temperature electrical resistance was recorded at different propane concentrations: 0%, 1%, 10%, 50%, 100%, 200%, 300%, 400% and 500%.

In order to study sensors sensitivity as a function of film thickness, sensitivity was obtained through the following equation:

$$S = \frac{|R - R_o|}{R_o} \tag{1}$$

Where S is expressed in percentage, R is the electrical resistance of the film in presence of propane gas and R_0 is the resistance registered at 0% of propane concentration.

3. Results and discussion

3.1 Film thickness

Table 1 presents the results of film thickness measurements obtained by both optical and mechanical profilometry. In some cases, film thickness could not be measured by mechanical profilometry because of a substrate deformation, attributed to mechanical stress generated after the annealing process. However, the values obtained are in the same range as those obtained by optical profilometry (20–150 nm); thus, giving support to the values obtained optically. Such support is needed since optically obtained values are generated from a theoretical model of the reflectance spectra generated during tests; these values range from 23 to 134 nm and have a correlation coefficient above 0.99, indicating an appropriate modeling and determination of film thickness. Because of this, the values obtained optically are used in the following analyses.

When plotting film thickness of TiO_2 as a function of immersion cycles an increasing trend with a linear behavior is obtained (**Figure 1**). The values varied from 22 nm to 130 nm and the linear fit yields a deposition rate of 18 nm per

Number immersions	Profilometry (nm)	Filmetrics (nm)
1	—	23.73
2	48.6	49.80
3	87.0	60.74
4	137.4	74.32
5	142.9	108.40
6	150.0	112.00
7	—	133.80

Table 1.
Results of the film thickness of TiO_2 samples, measured by profilometry and optical methods, as a function of the number of immersion cycles.

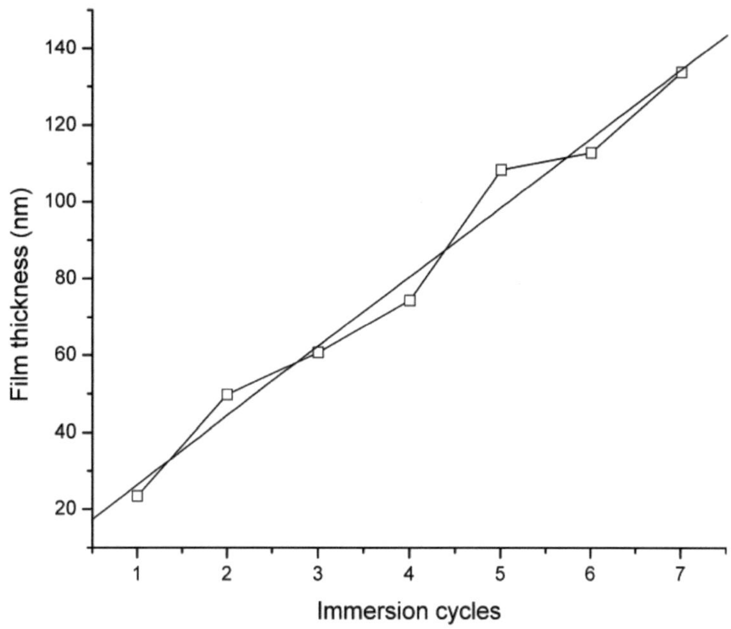

Figure 1.
Relation between the film thickness and the number of immersion cycles. A linear trend is obtained with a growing rate of 18 nm/cycle.

immersion cycle. When comparing our results to those reported by Hossein [21], where TiO_2 films were grown by the same method, a difference in film thickness is observed. Such difference is attributed to a larger solution molarity used by Hossein (0.4 M), that implies a larger number of particles available at the interphase for film growing.

It is well known that the electrical resistance of chemical gas sensors changes because of the exchange of charge carriers between the film surface and the chemisorbed species. Since the surface to volume ratio increases as the film thickness decreases, the amount of chemisorbed species, as well as the number of exchanged charge carriers, becomes larger than the number of charge carriers produced intrinsically in the bulk. At this point the output signal of the sensors becomes a function of the film thickness. Due to this effect and to the great control of film thickness obtained by this method, it could be possible to produce sensors with an optimized response. Considering that film thickness also depends on the solution molarity, which is due to the larger number of particles available close to the substrate surface, it is left for further work to analyze sensors response as a function of this parameter in order to optimize the deposition conditions.

3.2 X-ray diffraction

Figure 2 shows the X-ray spectra obtained for selected samples. In all spectra it is observed a diffraction peak at $2\theta = 25.23°$ indicating the presence of a crystalline structure corresponding to (101) planes of TiO_2 anatase structure, according to reported data in PDF 21-1272. Diffraction peaks corresponding to other titanium dioxide phases were not obtained. Given the deposition process, lattice distortions could be expected; however, from the lack of displacement of the diffraction peaks it can be inferred that there are no such distortions. This indicates that the temperature and annealing time were adequate. Such results agree with others reported elsewhere, where the anatase structure is obtained when heating TiO_2 films at 400° [33–35].

3.3 Surface morphology

2D and 3D atomic force micrographs of samples growth with one, three and five immersion cycles are displayed in **Figure 3**. From these figures it can be observed that the film growth becomes more compact and densely packed as the number of immersion cycles is increased; being the average grain size measured from the sample presented in **Figure 3e** and **f** of 50 ± 10 nm. Such differences in surface structure are of great importance for sensors behavior, since they are correlated to the exposed area, the number of available chemisorption sites, and the change in electrical resistance. The compactness of films also implies a better charge carrier transport because of the increased contact surface and the reduced spacing between grains. The observed surface structure will be used in the next section in order to explain the sensors behavior.

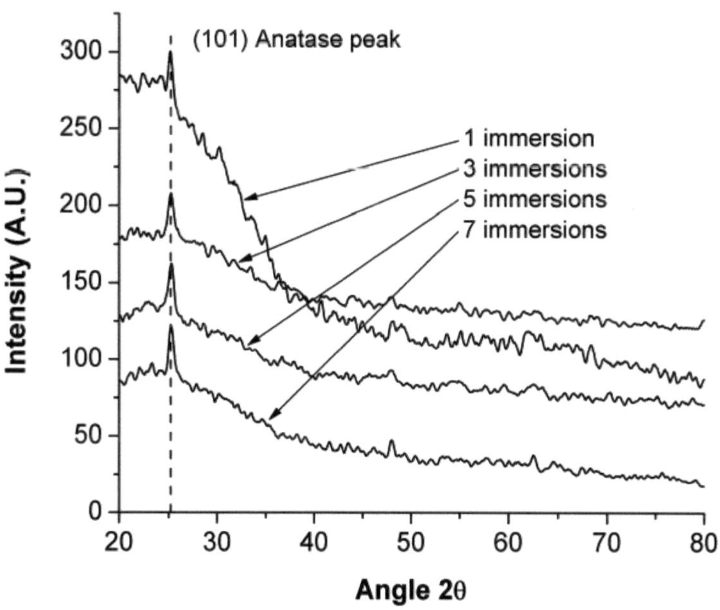

Figure 2.
XRD patterns from samples with different number of inmersions.

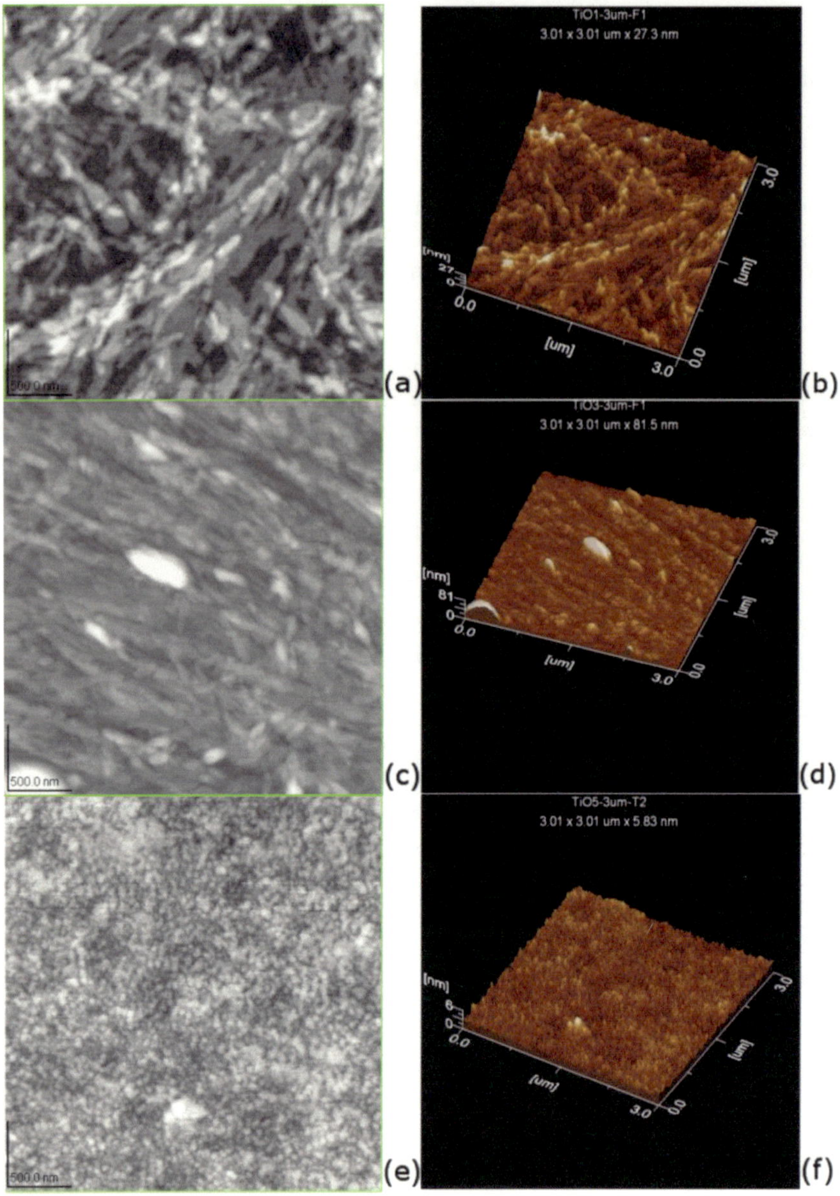

Figure 3.
2D and 3D AFM micrographs of TiO_2 thin films with one (a, b); three (c, d) and five immersion (e, f). cycles.

3.4 Sensing properties

Figure 4 shows graphs corresponding to samples grown with 1, 3, 5, and 7 immersions. In these graphs resistance versus gas concentration is plotted at different work temperatures. When analyzing the results it is observed that at temperatures of 200°C and below, changes in electrical resistance due to an increase of propane gas concentration are negligible and it can be considered as constant; however, when working at a temperature of 300°C all samples present important changes in electrical resistance, whose differences can be of an order of magnitude. As can be seen from all graphs, when temperature changes from 200 to 300°C, a drop in electrical resistance is obtained at a propane concentration of 0%; such drop is due to a thermal activation of electrical conductivity where a large number

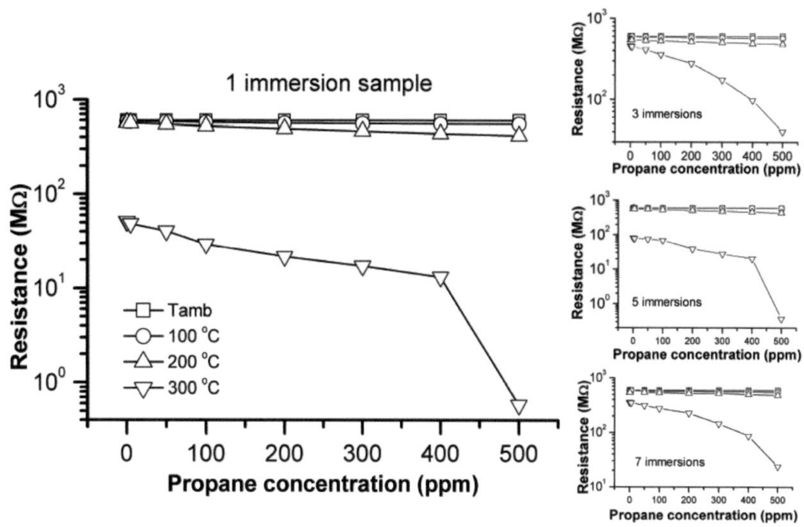

Figure 4.
Resistance as a function of propane concentration for different work temperatures. At 300°C changes in resistance become important. Inserted charts show the same behavior in all sensors.

of electrons are released into the conduction band. Further changes in electrical resistance are due to the interaction of propane gas with the films surface. Since all samples present the same behavior, the work temperature of 300°C will be used in the following analyses; however, it will be left for further analyses to find the minimum temperature at which the drop occurs, in order to minimize sensors power consumption.

Based on the n-type conductivity reported for the anatase structure of TiO_2 [36], the reduction of the electrical resistance of the films when they are in presence of propane can be attributed to an injection of electrons in the conduction band. This can be explained by the detection mechanism proposed by Kerlau [37], based on the studies of K. Cheng et al. [38], where propane molecules interact with lattice oxygen (O*) to form isopropoxyde and H-O* bonds, as it is explained by the following equation:

$$C_3H_8 + 2O^* \rightarrow C_3H_7O^* + H - O^* \quad (2)$$

Where the partial reaction between hydrogen and the lattice oxygen produces an extra electron that becomes trapped by the positive neighboring titanium atoms, and it is released into the conduction band by thermal excitation:

$$H + O^* + Ti^{4+} \rightarrow \left(H - O^*\right) + Ti^{3+} \quad (3)$$

The released electron contributes to increase the electrical conductivity, since it is a function of the number of free charge carriers, thus, diminishing its electrical resistance.

Figure 5a shows graphs of the sensitivity, as defined in Eq. (1), as a function of propane concentration in parts per million (ppm). As can be seen, in the range that goes from 5 to 300 ppm the sensor produced with five immersion cycles present the highest sensitivity and at higher propane concentrations sensors sensitivity tend to have the same values. A different perspective of this behavior is presented in **Figure 5b** where the sensitivity at the same propane concentration is plotted versus the number of immersion cycles. In this figure it is observed a dependence of the

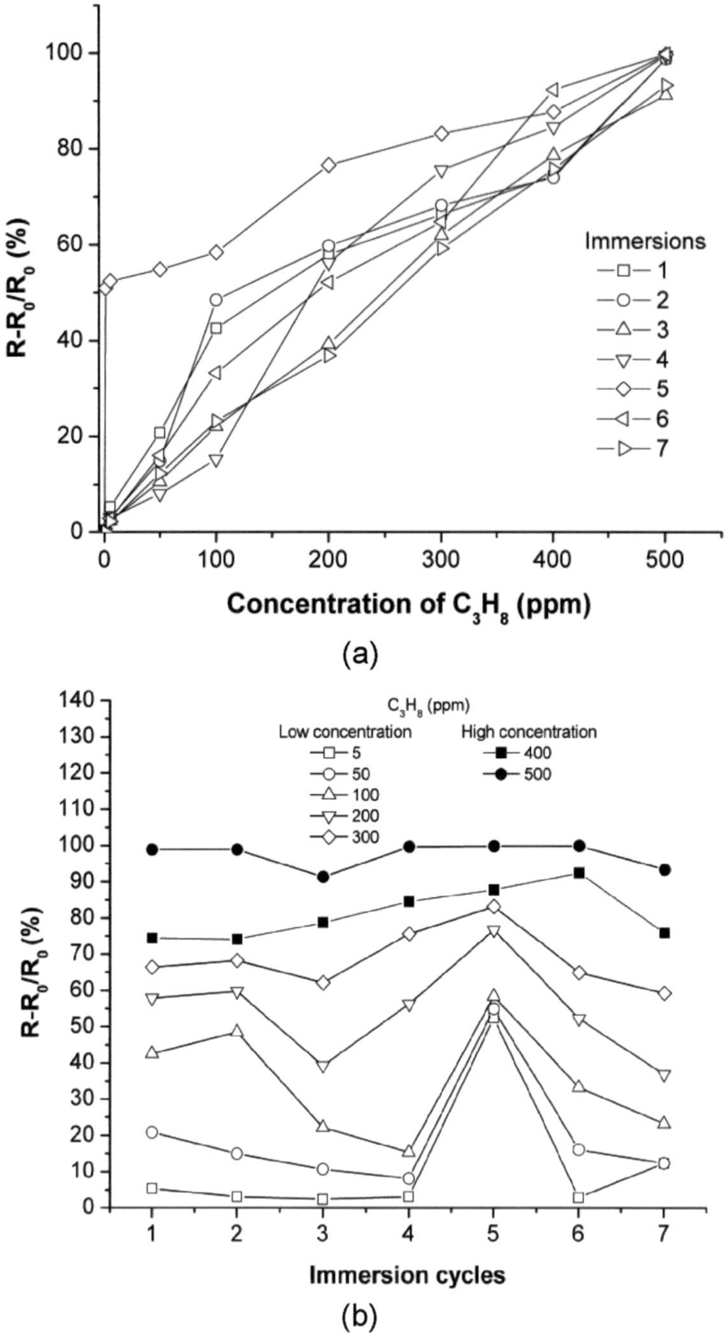

Figure 5.
Sensors sensitivity for the different films produced as a function of (a) propane concentrations and (b) as a function of the number of immersion cycles at different propane concentrations.

sensitivity not only as a function of immersion cycles (film thickness), but also as a function of propane concentration.

Based on the last, a two-regime behavior can be proposed: one at concentrations in the range that goes from 5 to 300 ppm and the other at concentrations above 300 ppm. In the first regime, or the low concentration regime, the values of sensitivity present a strong dependence of film thickness: a diminishing trend at a film thickness smaller than 60 nm, a maximum peak around 90 nm, and another diminishing trend at higher film thickness. In the second regime, or the regime of high concentrations, the sensitivity depends no more of the film thickness and presents a constant value for different film thickness.

The low concentration regime might be explained by the presence of a space charge region (SCR), generated by adsorbed particles at the sensor surface, and the surface morphology of the samples. Since positive hydrogen atoms are adsorbed at the surface, their electric fields bend the electronic state energies of titanium oxide towards lower energies (**Figure 6**). This band bending is projected into the material and becomes less important as it goes deeper into the film. This is due to the decaying electric field tendency, which means that the energy E of free carriers in the conduction band changes with depth as:

$$E = E_C - \frac{A}{r^n} \tag{4}$$

Where E_c is the energy at the edge of the conduction band, A and n are constants that depend on crystal structure and number of chemisorbed particles, and r is the distance from the surface. If the film thickness is small enough to be comparable in magnitude to the depth of the SCR, the contribution of the SCR to the electrical conductivity of the film becomes dominant and depends directly on the number of adsorbed atoms.

Based on this, it is expected a diminishing trend of sensitivity with immersion cycles, for every cycle increases films thickness and reduces the contribution to electrical conductivity of the SCR. This explains the behavior up to 70 nm (4 immersion cycles); however, it does not explain the behavior at larger film thickness. The peak at 5 immersion cycles followed by a diminishing trend in sensitivity is attributed to a combined effect of the SCR and changes in surface structure. According to AFM images, the compactness of films with larger thickness and their more uniform distribution of grains result in a better electrical contact among crystallites, not only in the bulk but also at the surface. Since electrical conduction is given in the surface due to the SCR, a more uniform surface allows a larger amount of charges at the interphase and a conduction path with an inferior amount of dislocations and vacancies contributing to diminish carriers' mobility. Since the change in surface structure allows a more adequate interaction of the sensor with the gas it is expected an increase in sensors sensitivity, and, thus, the peak observed at five cycles. This

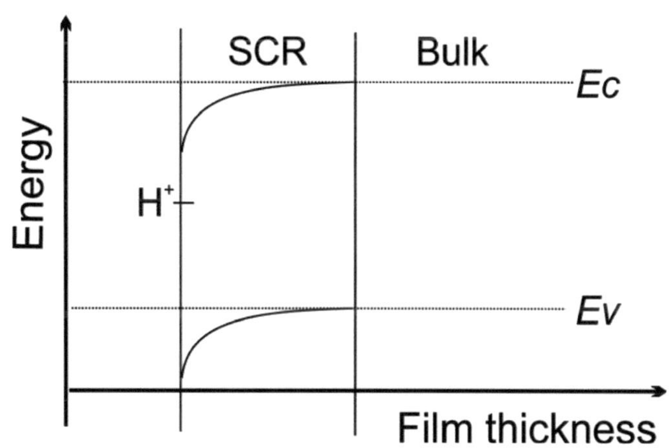

Figure 6.
Energy band diagram as a function of the film thickness. The chemisorbed proton H^+ bends the energy bands towards lower energies, this bending known as the space charge region (SCR), becomes less important as it goes deeper in the film; at large film thickness the band gap equals that of the bulk.

behavior is given in samples with a larger thickness than 110 nm but, since the influence of the SCR diminishes with the thickness, the diminishing trend is maintained for a larger number of immersion cycles.

The high concentration regime, where sensitivity becomes independent of film thickness is explained by the saturation of the surface by the adsorbed species. According to Langmuir theory the number of adsorbed species is a function of gas partial pressure. This means that at high propane concentrations the number of adsorbed species is large enough that the contribution to the electrical conductivity due to differences in surface area is negligible. In this regime sensitivity depends only on propane partial pressure.

4. Conclusions

Seven titanium oxide gas sensors were produced on glass substrates by the ultrasonically assisted dip-coating method, using a 0.17 M solution of titanium isopropoxide and an immersion speed of 0.122 cm/min. Film thicknesses ranged from 20 nm to 150 nm with deposition rate of 18 nm per immersion. From X-ray diffraction analysis, sensors annealed at 400°C for 3 hrs a polycrystalline structure corresponding to the TiO_2 anatase phase.

Propane sensing tests indicate that work temperatures of 300°C cause significant changes in electrical resistance of TiO2 gas sensors and is close to be the smallest temperature at which sensors can detect propane presence. An analysis of the sensitivity as a function of the film thickness indicates a two-regime behavior where at concentrations above 400 ppm of propane the sensitivity becomes independent of the thickness. Sensitivity becomes dependent of film thickness at lower concentrations. Its behavior is explained by a contribution of a space charge region and a change in surface structure. Such behavior indicates that film thickness values close to 120 nm are adequate to sense propane at concentrations in the range of 5 to 300 ppm of propane.

Acknowledgements

The authors thank and recognize the financial support of DGAPA-UNAM Project IN102419. Useful discussions with of Dr. Carlos Magaña and Dr. Francisco Hernández are recognized and appreciated. Also we thank the technical assistance of Manuel Aguilar in laboratory works.

DOI: http://dx.doi.org/10.5772/intechopen.95528

Author details

Dwight Acosta-Najarro[1]*, Ismael A. Garduño-Wilches[2], Maria de la Luz Olvera[3] and Arturo Maldonado[3]

1 Instituto de Física, Universidad Nacional Autónoma de México, Mexico

2 Cátedras Conacyt/Instituto Politécnico Nacional, Centro de Investigación en Ciencia Aplicada y Tecnología Avanzada, Calzada Legaria 694, Col. Irrigación, Del. Miguel Hidalgo, C. P. 11500, Ciudad de México, Mexico

3 SEES, Centro de Investigacion y de Estudios Avanzados del Instituto Politécnico Nacional, 14740, Mexico City, Mexico

*Address all correspondence to: dacosta@fisica.unam.mx

References

[1] J.A. Papa, Propanols, Ulmann's Encycl. Ind. Chem. (2000).

[2] B. Bahavar, M.I. Chaudhiy, R.J. McCluskey, Hetero-Epitaxial Growth of SiC Films by CVD from Silane, Methane, Propane, and Hydrogen Mixtures, MRS Online Proc. Libr. 242 (1992). doi:10.1557/PROC-242-555.

[3] D. Rathee, M. Kumar, S.K. Arya, Deposition of nanocrystalline thin TiO2 films for MOS capacitors using Sol–Gel spin method with Pt and Al top electrodes, Solid. State. Electron. 76 (2012) 71-76. doi:10.1016/j.sse.2012.04.041.

[4] R.L. Porto, R. Frappier, J.B. Ducros, C. Aucher, H. Mosqueda, S. Chenu, et al., Titanium and vanadium oxynitride powders as pseudo-capacitive materials for electrochemical capacitors, Electrochim. Acta. 82 (2012) 257-262. doi:10.1016/j.electacta.2012.05.032.

[5] K.F. Albertin, I. Pereyra, Study of reactive sputtering titanium oxide for metal-oxide-semiconductor capacitors, Thin Solid Films. 517 (2009) 4548-4554. doi:10.1016/j.tsf.2008.12.045.

[6] D. Li, H. Haneda, S. Hishita, N. Ohashi, Visible-light-driven nitrogen-doped TiO 2 photocatalysts : effect of nitrogen precursors on their photocatalysis for decomposition of gas-phase organic pollutants, Mater. Sci. Eng. B. 117 (2005) 67-75. doi:10.1016/j.mseb.2004.10.018.

[7] Y. Chen, F. Chen, J. Zhang, Effect of surface fluorination on the photocatalytic and photo-induced hydrophilic properties of porous TiO2 films, Appl. Surf. Sci. 255 (2009) 6290-6296. doi:10.1016/j.apsusc.2009.02.004.

[8] E.M. El-Maghraby, Y. Nakamura, S. Rengakuji, Composite TiO2–SnO2 nanostructured films prepared by spin-coating with high photocatalytic performance, Catal. Commun. 9 (2008) 2357-2360. doi:10.1016/j.catcom.2008.05.037.

[9] M. a Arenas, C. Pérez-Jorge, A. Conde, E. Matykina, J.M. Hernández-López, R. Pérez-Tanoira, et al., Doped TiO(2) anodic layers of enhanced antibacterial properties., Colloids Surf. B. Biointerfaces. 105C (2013) 106-112. doi:10.1016/j.colsurfb.2012.12.051.

[10] M. Senna, N. Myers, A. Aimable, V. Laporte, C. Pulgarin, O. Baghriche, et al., Modification of titania nanoparticles for photocatalytic antibacterial activity via a colloidal route with glycine and subsequent annealing, J. Mater. Res. 28 (2012) 354-361. doi:10.1557/jmr.2012.257.

[11] J. a. Dirksen, K. Duval, T. a. Ring, NiO thin-film formaldehyde gas sensor, Sensors Actuators B Chem. 80 (2001) 106-115. doi:10.1016/S0925-4005(01)00898-X.

[12] K. Galatsis, Y.X. Li, W. Wlodarski, E. Comini, G. Sberveglieri, C. Cantalini, et al., Comparison of single and binary oxide MoO3, TiO2 and WO3 sol–gel gas sensors, Sensors Actuators B Chem. 83 (2002) 276-280. doi:10.1016/S0925-4005(01)01072-3.

[13] D. Koziej, N. Bârsan, V. Hoffmann, J. Szuber, U. Weimar, Complementary phenomenological and spectroscopic studies of propane sensing with tin dioxide based sensors, Sensors Actuators B Chem. 108 (2005) 75-83. doi:10.1016/j.snb.2004.12.076.

[14] R.H.R. Castro, P. Hidalgo, H.E.M. Perez, F.J. Ramirez-Fernandez, D. Gouvêa, Relationship between surface segregation and rapid propane electrical response in Cd-doped SnO2

nanomaterials, Sensors Actuators B Chem. 133 (2008) 263-269. doi:10.1016/j.snb.2008.02.021.

[15] S.N. Oliaee, A. Khodadadi, Y. Mortazavi, S. Alipour, Highly selective Pt/SnO2 sensor to propane or methane in presence of CO and ethanol, using gold nanoparticles on Fe2O3 catalytic filter, Sensors Actuators B Chem. 147 (2010) 400-405. doi:10.1016/j.snb.2010.03.061.

[16] P. Dutta, M. Frank, G. Hunter, M. George, Reactively sputtered titania films as high temperature carbon monoxide sensors, Sensors Actuators B Chem. 106 (2005) 810-815. doi:10.1016/j.snb.2004.09.036.

[17] Z.M. Seeley, A. Bandyopadhyay, S. Bose, Titanium dioxide thin films for high temperature gas sensors, Thin Solid Films. 519 (2010) 434-438. doi:10.1016/j.tsf.2010.07.040.

[18] S. Rahbarpour, S.M. Hosseini-Golgoo, Diode type Ag–TiO2 hydrogen sensors, Sensors Actuators B Chem. 187 (2013) 262-266. doi:10.1016/j.snb.2012.11.017.

[19] Y. Kimura, S. Kimura, R. Kojima, M. Bitoh, M. Abe, M. Niwano, Micro-scaled hydrogen gas sensors with patterned anodic titanium oxide nanotube film, Sensors Actuators B Chem. 177 (2013) 1156-1160. doi:10.1016/j.snb.2012.12.016.

[20] F. Hossein-Babaei, M. Keshmiri, M. Kakavand, T. Troczynski, A resistive gas sensor based on undoped p-type anatase, Sensors Actuators B Chem. 110 (2005) 28-35. doi:10.1016/j.snb.2005.01.010.

[21] M.Z. Atashbar, Nano-sized TiO_2 thin film for alcohol sensingapplication, Proc. 2001 1st IEEE Conf. Nanotechnology. IEEE-NANO 2001 (Cat. No.01EX516). (2001) 544-549. doi:10.1109/NANO.2001.966482.

[22] T. Toccoli, S. Capone, L. Guerini, M. Anderle, A. Boschetti, E. Iacob, et al., Growth of titanium dioxide films by cluster supersonic beams for VOC sensing applications, IEEE Sens. J. 3 (2003) 199-205. doi:10.1109/JSEN.2003.812627.

[23] N. Verma, S. Singh, R. Srivastava, B.C. Yadav, Fabrication of iron titanium oxide thin film and its application as opto-electronic humidity and liquefied petroleum gas sensors, Opt. Laser Technol. 57 (2014) 181-188. doi:10.1016/j.optlastec.2013.10.007.

[24] T.S. Yang, M.C. Yang, C. Bin Shiu, W.K. Chang, M.S. Wong, Effect of N 2 ion flux on the photocatalysis of nitrogen-doped titanium oxide films by electron-beam evaporation, Appl. Surf. Sci. 252 (2006) 3729-3736. doi:10.1016/j.apsusc.2005.05.070.

[25] J.S. Kim, H. Jee, Y.H. Yu, H.W. Seo, Titanium dioxide and suboxide thin films grown with controlled discharge voltage in reactive direct-current sputtering, Thin Solid Films. 672 (2019) 14-21. doi:10.1016/j.tsf.2018.12.045.

[26] H.C. Hsieh, J. Yu, S.P. Rwei, K.F. Lin, Y.C. Shih, L. Wang, Ultra-compact titanium oxide prepared by ultrasonic spray pyrolysis method for planar heterojunction perovskite hybrid solar cells, Thin Solid Films. 659 (2018) 41-47. doi:10.1016/j.tsf.2018.05.002.

[27] F. Rahmani, M. Ardyanian, Fabrication and characterization of ZnO/TiO2 multilayers, deposited via spin coating method, J. Mater. Sci. Mater. Electron. 29 (2018) 4285-4293. doi:10.1007/s10854-017-8375-3.

[28] R. De Aguiar Ramos, M.H. Boratto, M.S. Li, L.V. De Andrade Scalvi, Emission properties related to distinct phases of sol-gel dip-coating titanium dioxide, and carrier photo-excitation in different energy

ranges, Mater. Res. 20 (2017) 866-873. doi:10.1590/1980-5373-MR-2016-0675.

[29] G. Govindasamy, P. Murugasen, S. Sagadevan, Investigations on the Synthesis, Optical and Electrical Properties of TiO2 Thin Films by Chemical Bath Deposition (CBD) method, Mater. Res. 19 (2016) 413-419. doi:10.1590/1980-5373-MR-2015-0411.

[30] D. V. Pinjari, K. Prasad, P.R. Gogate, S.T. Mhaske, A.B. Pandit, Synthesis of titanium dioxide by ultrasound assisted sol-gel technique: Effect of calcination and sonication time, Ultrason. Sonochem. 23 (2015) 185-191. doi:10.1016/j.ultsonch.2014.10.017.

[31] S. Arunmetha, P. Manivasakan, A. Karthik, N.R. Dhinesh Babu, S.R. Srither, V. Rajendran, Effect of processing methods on physicochemical properties of titania nanoparticles produced from natural rutile sand, Adv. Powder Technol. 24 (2013) 972-979. doi:10.1016/j.apt.2013.01.011.

[32] J. Aguilar-Leyva, A. Maldonado, M. de la L. Olvera, Gas-sensing characteristics of undoped-SnO2 thin films and Ag/SnO2 and SnO2/Ag structures in a propane atmosphere, Mater. Charact. 58 (2007) 740-744. doi:10.1016/j.matchar.2006.11.016.

[33] a. Elfanaoui, E. Elhamri, L. Boulkaddat, a. Ihlal, K. Bouabid, L. Laanab, et al., Optical and structural properties of TiO2 thin films prepared by sol–gel spin coating, Int. J. Hydrogen Energy. 36 (2011) 4130-4133. doi:10.1016/j.ijhydene.2010.07.057.

[34] S. Mahshid, M.S. Ghamsari, M. Askari, N. Afshar, S. Lahuti, Synthesis of TiO 2 nanoparticles by hydrolysis and peptization of titanium isopropoxide solution, Semicond. Physics, Quantum Electron. Optoelectron. 9 (2006) 65-68.

[35] M.M. Hasan, A.S.M.A. Haseeb, R. Saidur, H.H. Masjuki, Effects of Annealing Treatment on Optical Properties of Anatase TiO 2 Thin Films, (2008) 92-96.

[36] N. Savage, B. Chwieroth, A. Ginwalla, B.R. Patton, S.A. Akbar, P.K. Dutta, Composite n–p semiconducting titanium oxides as gas sensors, Sensors Actuators B Chem. 79 (2001) 17-27. doi:10.1016/S0925-4005(01)00843-7.

[37] M. Kerlau, P. Reichel, N. Bârsan, U. Weimar, S. Delsarte-Guéguen, O. Merdrignac-Conanec, Detection of propane by "GaON" thick-film gas sensors, Sensors Actuators B Chem. 122 (2007) 14-19. doi:10.1016/j.snb.2006.05.001.

[38] K. Chen, E. Iglesia, A.T. Bell, Kinetic Isotopic Effects in Oxidative Dehydrogenation of Propane on Vanadium Oxide Catalysts, J. Catal. 192 (2000) 197-203. doi:10.1006/jcat.2000.2832.